INTERNATIONAL ENCYCLOPAEDIA OF GENETICS, GENOMICS AND PROTEOMICS

INTERNATIONAL ENCYCLOPAEDIA OF GENETICS, GENOMICS AND PROTEOMICS

B.G. Mukherjee

ANMOL PUBLICATIONS PVT. LTD.
NEW DELHI - 110 002 (INDIA)

ANMOL PUBLICATIONS PVT. LTD.
H.O.: 4374/4B, Ansari Road, Daryaganj,
New Delhi-110 002 (India)
Ph.: 23278000, 23261597
B.O.: No. 1015, Ist Main Road, BSK IIIrd Stage
IIIrd Phase, IIIrd Block,
Bangalore - 560 085 (India)
Visit us at: www.anmolpublications.com

International Encyclopaedia of Genetics, Genomics and Proteomics

ISBN 978-81-261-3592-9

PRINTED IN INDIA

Printed at Mehra Offset Press, Delhi.

Contents

Code, Gene Expression and Evolution; Overview of Major Genetic Research Organizations; Genome and Genomics: Major Areas of Study and Project; Science of Proteomics: Major Dimensions and Trends; Glossary etc.

—Editor

Preface

This publication titled "International Encyclopaedia of Genetics, Genomics and Proteomics" deals with the subject in its entirity. Genetics is the scientific study of heredity and addresses questions such as, how particular qualities or traits are transmitted from parents to offspring. Genomics refers to mapping, sequencing and analysis of the entire DNA sequence of an organism or species. DNA (Deoxyribonucleic Acid) is the molecule that carries the genetic information for most living systems.

The DNA molecule consists of four bases (adenine, cytosine, guanine and thymine) and a sugar phosphate backbone arranged in two connected strands. Proteomics as a term has come to have a very broad meaning including the manipulation and alteration of the genetic material (constitution) of an organism in such a way as to allow it to produce endogenous proteins with properties different from those of the traditional (historic/typical), or to produce entirely different (foreign) proteins altogether.

Some other words often applicable to the same process are gene splicing, gene manipulation, or recombinant DNA technology (techniques). Today, all these are a rapidly growing fields of research and development that is cutting across many traditional boundaries. This encyclopaedia titled "International Encyclopaedia of Genetics & Genomics & Proteomics " is really unique in terms of coverage, scope, elaboration, usefulness and futuristic perspective.

It will provide readers with a user-friendly approach towards acquainting themselves with various select traditional and newly emerging fields of biosciences. This encyclopaedia is of great contemporary relevance. It will serve as a reference book on the said subject and tends to acquaint readers with both in-depth and extensive knowledge of the subject concerned.

This encyclopaedia is a unique initiative to provide the most comprehensive coverage to the subject area in question. This encyclopedia mainly focuses on the following areas: Introduction to Genetics and Geneticists; Elements of Genetic

1

Introduction to Genetics & Geneticists

GENETICS

Genetics is the science of heredity and variation in living organisms. Knowledge of the inheritance of characteristics has been implicitly used since prehistoric times for improving crop plants and animals through selective breeding. However, the modern science of genetics, which seeks to understand the mechanisms of inheritance, only began with the work of Gregor Mendel in the mid-1800s. Although he did not know the physical basis for heredity, Mendel observed that inheritance is fundamentally a discrete process with specific traits that are inherited in an independent manner — these basic units of inheritance are now called genes.

Following the rediscovery of Gregor Johann Mendel's observations in the early 1900s, research in 1910s yielded the first physical understanding of inheritance — that genes are arranged linearly along large cellular structures called chromosomes. By the 1950s it was understood that the core of a chromosome was a long molecule called DNA and genes existed as linear sections within the molecule. A single strand of DNA is a chain of four types of nucleotides; hereditary information is contained within the sequence of these nucleotides. Solved by Watson, Wilkins, and Crick in 1953, DNA's three-dimensional structure is a double-stranded helix, with the nucleotides on each strand physically matched to each other. Each strand acts as a template for synthesis of a new partner strand, providing the physical mechanism for the inheritance of information.

The sequence of nucleotides in DNA is used to produce specific sequences of amino acids, creating proteins — a correspondence known as the "genetic code". This sequence of amino acids in a protein determines how it folds into a three-dimensional structure, this structure is in turn responsible for the protein's function.

Proteins are responsible for almost all functional roles in the cell. A change to DNA sequence can change a protein's structure and behavior, and this can have dramatic consequences in the cell and on the organism as a whole.

Although genetics plays a large role in determining the appearance and behavior of organisms, it is the interaction of genetics with the environment an organism experiences that determines the ultimate outcome. For example, while genes play a role in determining a person's height, the nutrition and health that person experiences in childhood also have a large effect.

History of Genetics

Although the science of genetics has its origins in the work of Gregor Mendel in the mid-1800s, various theories of inheritance preceded Mendel. These theories generally assumed that there existed an inheritance of acquired characteristics (also known as "soft inheritance"): the belief that individuals inherit traits that have been strengthened in their parents. Today, the theory is commonly associated with Jean-Baptiste Lamarck, who used this pattern of inheritance to explain the evolution of various traits within species (these changes are now understood to be the product of natural selection rather than a product of soft inheritance).

Mendelian and Classical Genetics

The modern science of genetics traces its roots to the observations made by Gregor Johann Mendel, a German-Czech Augustinian monk and scientist who made detailed studies of the nature of inheritance in plants. In his paper "Versuche über Pflanzenhybriden" ("Experiments on Plant Hybridization"), presented in 1865 to the Brunn Natural History Society, Gregor Mendel traced the inheritance patterns of certain traits in pea plants and showed that they could be described mathematically. Although not all features show these patterns of Mendelian inheritance, his work suggested the utility of the application of statistics to the study of inheritance.

The significance of Mendel's observations was not understood until early in the twentieth century, after his death, when his research was re-discovered by other scientists working on similar problems. The word "genetics" itself was coined in 1905 by William Bateson, a significant proponent of Mendel's work, in a letter to Adam Sedgwick. The adjective "genetic" (derived from the Greek word "genno" to give birth) predates the noun, dating back to the 1830's and first used in the biological sense in 1859 by Charles Darwin in the *The Origin of Species*. Bateson publicly promoted and popularized usage of word "genetics" to describe the study of inheritance in his inaugural address to the Third International Conference on Plant Hybridization in London, England, in 1906.

In the decades following rediscovery and popularization of Mendel's work, numerous experiments sought to elucidate the molecular basis of DNA. In 1910 Thomas Hunt Morgan argued that genes reside on chromosomes, based on observations of a sex-linked white eye mutation in fruit flies. In 1913 his student Alfred Sturtevant used the phenomenon of genetic linkage and the associated recombination rates to demonstrate and map the linear arrangement of genes upon the chromosome.

Molecular Genetics

Although chromosomes were known to contain genes, chromosomes were composed of both protein and DNA — it was unknown which was critical for heredity or how the process occurred. In 1928, Frederick Griffith published his discovery of the phenomenon of transformation (see Griffith's experiment); sixteen years later, in 1944, Oswald Theodore Avery, Colin McLeod and Maclyn McCarty used this phenomenon to isolate and identify the molecule responsible for transformation as DNA. The Hershey-Chase experiment in 1952 identified DNA (rather than protein) as the genetic material of viruses, further evidence that DNA was the molecule responsible for inheritance.

James D. Watson and Francis Crick resolved the structure of DNA in 1953, using the X-ray crystallography work of Rosalind Franklin that indicated the molecule had a helical structure. Their double-helix model paired a sequence of nucleotides with a "complement" on the other strand. This structure not only provided a physical explanation for information contained within the order of the nucleotides, but also a physical mechanism for duplication through separation of strands and the reconstruction of a partner strand based on the nucleotide pairings. Although the structure explained the process of inheritance, it was still unknown how DNA influenced the behavior of cells. In the following years many scientists sought to understand how DNA controls the process of protein production within ribosomes, eventually discovering the transcription of DNA into messenger RNA and uncovering the genetic code which links the nucleotide sequence of messenger RNA to the amino acid sequence of protein.

With this molecular understanding of DNA, an explosion of research based on this understanding of the molecular nature of DNA became possible. The development of chain-termination DNA sequencing in 1977 enabled the determination of nucleotide sequences on DNA, and the PCR method developed by Kary Banks Mullis in 1983 allowed the isolation and amplification of arbitrary segments of DNA. These and other techniques, through the pooled efforts of the Human Genome Project and parallel private effort by Celera Genomics, culminated in the sequencing of the human genome in 2001.

Features of Inheritance

Discrete Inheritance and Mendel's Laws

At its most fundamental level, inheritance in organisms occurs by means of discrete traits, called "genes". This property was first observed by Gregor Mendel, who studied the segregation of heritable traits in pea plants. In his experiments studying the trait for flower color, Mendel observed that the flowers of each pea plant were either purple or white — and never an intermediate between the two colors. These different, discrete versions of the same gene are called "alleles".

In the case of pea plants, each organism has two alleles of each gene, and the plants inherit one allele from each parent. Many organisms, including humans, have this pattern of inheritance. Organisms with two copies of the same allele are called "homozygous", while organisms with two different alleles are "heterozygous".

The set of alleles for a given organism is called its genotype, while the visible trait the organism has is called its "phenotype". When organisms are heterozygous, often one allele is called "dominant" as its qualities "dominate" the phenotype of the organism, while the other allele is called "recessive" as its qualities "recede" and are not observed. Dominant alleles are often abbreviated with a capital letter, while recessive alleles are given a lowercase version of the same letter. Some alleles do not have complete dominance and instead have incomplete dominance by expressing an intermediate phenotype, or codominance by expressing both alleles at once.

When parents breed to produce children, their children randomly inherit one of the two alleles from each parent. The outcome of these crosses can be visualized by use of a Punnett square. These observations of discrete inheritance and the segregation of alleles are collectively known as "Mendel's first law" or the "Law of Segregation".

Assortment and Interactions of Multiple Genes

Organisms have thousands of genes, and in sexually reproducing organisms assortment of these genes are generally independent of each other. This means that the inheritance of an allele for yellow or green pea color is unrelated to the inheritance of alleles for white or purple flowers. This phenomenon, known as "Mendel's second law" or the "Law of independent assortment", means that the alleles of different genes get shuffled between parents to form children with many different combinations. (Some genes do not assort independently, demonstrating genetic linkage, a topic discussed later in this article.)

Often different genes can interact in a way that influences the same trait. In the blue-eyed Mary, for example, there exists a gene with alleles that determine the color of flowers: blue or magenta. Another gene, however, controls whether the flowers have color at all: color or white. When a plant has two copies of this white allele, its flowers are white — regardless of whether the first gene has blue or magenta alleles. This interaction between genes is called "epistasis", with the second gene epistatic to the first.

Many traits are not discrete features (eg. purple or white flowers) but are instead continuous features (eg. human height and skin color). These "complex traits" are the product of interactions of many genes. The influence of these genes is mediated, to varying degrees, by the environment an organism has experienced. The degree to which an organism's genes contribute to a complex trait is called "heritability". Measurement of the heritability of a trait is relative, though — in a more variable environment, the environment has a bigger influence on the total variation of the trait. For example, human height is a complex trait with a heritability of 89% in the United States. In Nigeria, however, where people experience a more variable access to good nutrition and health care, height has a heritability of only 62%.

The Molecular Basis for Inheritance

DNA and Chromosomes

The molecular basis for genes is deoxyribonucleic acid (DNA). DNA is composed of a chain of nucleotides, of which there are four types: adenine (A), cytosine (C), guanine (G), and thymine (T). Genetic information exists in the sequence of these nucleotides, and genes exist as stretches of sequence along the DNA chain. Viruses are the only exception to this rule — sometimes viruses use the very similar molecule RNA instead of DNA as their genetic material. The full set of all hereditary material in an organism is called its "genome".

DNA normally exists as a double-stranded molecule, coiled into the shape of a double-helix. Each nucleotide in DNA preferentially pairs with its partner nucleotide on the opposite strand: A pairs with T, and C pairs with G. Thus, in its two-stranded form, each strand effectively contains all necessary information, redundant with its partner strand. This structure of DNA is the physical basis for inheritance: DNA replication duplicates the genetic information by splitting the strands and using each strand as a template for synthesis of a new partner strand.

Genes are arranged linearly along the long chains of DNA sequence, called chromosomes. In bacteria, each cell has a single circular chromosome, while eukaryotic organisms (which includes plants and animals) have their DNA arranged in multiple linear chromosomes. These DNA strands are often extremely long; the largest human chromosome, for example, is about 247 million base pairs in length. The DNA of a chromosome is associated with structural proteins which organize, compact, and control access to the DNA, forming a material called chromatin; in eukaryotes chromatin is usually composed of nucleosomes, repeating units of DNA wound around a core of histone proteins. The full set of hereditary material in an organism (usually the combined DNA sequences of all chromosomes) is called the "genome".

While haploid organisms have only one copy of each chromosome, most animals and many plants are diploid, containing two copies of each chromosome and thus two copies of every gene. The two alleles for a gene are located on identical loci of sister chromatids, each allele inherited from a different parent.

An exception exists in the sex chromosomes, specialized chromosomes many animals have evolved that play a role in determining the sex of the organism. In humans and other mammals the Y chromosome has very few genes and triggers the development of male sexual characteristics, while the X chromosome is similar to the other chromosomes and contains many genes unrelated to sex determination. Females have two copies of the X chromosome, but males have one Y and only one X chromosome — this difference in X chromosome copy numbers leads to the unusual inheritance patterns of sex linked disorders.

Reproduction

When cells divide, their full genome is copied and each daughter cell inherits one copy. This is a simplest form of reproduction and is the basis for asexual reproduction. Asexual reproduction can also occur in multicellular organisms, always producing children which each inherit a full copy of a single parental genome. When asexual reproduction occurs, the child organisms are "clones" as they contain the same genetic material as the parent.

Eukaryotic organisms often use sexual reproduction to generate children which contain a mixture of genetic material inherited from two different parents. The process of sexual reproduction generally alternates between forms which contain single copies of the genome (haploid) and double copies (diploid). Haploid cells fuse and combine genetic material to create a diploid cell with paired chromosomes. Diploid organisms form haploids by dividing, without replicating their DNA, to create daughter cells that randomly inherit one of each pair of chromosomes. Most animals and many plants are diploid for the majority of their lifespan, with the haploid form reduced to single cell gametes.

Although they do not use the haploid/diploid method of sexual reproduction, bacteria have many methods of acquiring new genetic information. Some bacteria can undergo conjugation, transferring a small circular piece of DNA to another bacterium. Bacteria can also take up raw DNA fragments found in the environment and integrate them into their genome, a phenomenon known as transformation. This processes result in horizontal gene transfer, transmitting fragments of genetic information between organisms that would otherwise be unrelated.

Genetic Recombination and Linkage

The diploid nature of chromosomes allows for genes on different chromosomes to assort independently during sexual reproduction, recombining to form new combinations of genes. Genes on the same chromosome would theoretically never recombine, however, were it not for the process of chromosomal crossover. During crossover, chromosomes exchange stretches of DNA, effectively shuffling the gene alleles between the chromosomes to . This process of chromosomal crossover generally occurs during meiosis, a series of cell divisions that creates haploid germ cells which later combine with other germ cells to form child organisms.

The probability of chromosomal crossover occurring between two given points on the chromosome is related to the distance between them. For an arbitrarily long distance, the probability of crossover is high enough that the inheritance of the genes is effectively uncorrelated. For genes that are closer together, however, the lower probability of crossover means that the genes demonstrate genetic linkage — alleles for the two genes tend to be inherited together. The amounts of linkage between a series of genes can be combined to form a linear linkage map that roughly describes the arrangement of the genes along the chromosome.

Gene Expression and the Creation of Phenotype

The Genetic Code

Genes generally express their functional effect through the production of proteins, which are complex molecules responsible for most functions in the cell. Proteins are chains of amino acids, and the DNA sequence of a gene (through an RNA intermediate) is used to produce a specific protein sequence. Each group of three nucleotides in the sequence, called a codon, corresponds to one of the twenty possible amino acids in protein — this correspondence is called the genetic code. The flow of information is unidirectional: information is transferred from nucleotide sequences into the amino acid sequence of proteins, but never from protein back into the sequence of DNA — a phenomenon Francis Crick called the central dogma of molecular biology.

The specific sequence of amino acids results in a unique three-dimensional structure for that protein, and the three-dimensional structures of protein are related to their function. Some are simple structural molecules, like the fibers formed by the protein collagen. Proteins can bind to other proteins and simple molecules, sometimes acting as enzymes by facilitating chemical reactions within the bound molecules (without changing the structure of the protein itself). Protein structure is dynamic; the protein hemoglobin bends into slightly different forms as it facilitates the capture, transport, and release of oxygen molecules within mammalian blood.

A single nucleotide difference within DNA can cause a single change in the amino acid sequence of a protein. Because protein structures are the result of their amino acid sequences, some changes can dramatically change the properties of a protein by destabilizing the structure or changing the surface of the protein in a way that changes its interaction with other proteins and molecules. For example, sickle-cell anemia is a human genetic disease that results from a single base difference within the coding region for the â-globin section of hemoglobin, causing a single amino acid change that changes hemoglobin's physical properties. Sickle-cell versions of hemoglobin stick to themselves, stacking to form fibers that distort the shape of red blood cells carrying the protein. These sickle-shaped cells no longer flow smoothly through blood vessels, having a tendency to clot or degrade, causing the medical problems associated with the disease.

Nature vs. Nurture

Although genes contain all the information an organism uses to function, the environment plays an important role in determining the ultimate phenotype — a dichotomy often referred to as "nature vs. nurture". The phenotype of an organism depends on the interaction of genetics with the environment. One example of this is the case of temperature-sensitive mutations. Often, a single amino acid change within the sequence of a protein does not change its behavior and interactions with other molecules, but it does destabilize the structure. In a high temperature environment, where molecules are moving more quickly and hitting each other, this results in the protein losing its structure and failing to function. In a low temperature environment, however, the protein's structure is stable and functions normally. This sort of mutation is visible in the coat coloration of Siamese cats, where a mutation in an enzyme responsible for pigment production causes it to destabilize and lose function at high temperatures. The protein remains functional in areas of skin that are colder — legs, ears, tail, and face — and so the cat has dark fur at its extremities.

Environment also plays a dramatic role in effects of the human genetic disease phenylketonuria. The mutation that causes phenylketonuria disrupts the ability of the body to break down the amino acid phenylalanine, causing toxic build-up of an intermediate molecule that, in turn, causes severe symptoms of progressive mental retardation and seizures. If someone with the phenylketonuria mutation is kept on a strict diet that avoids this amino acid, however, they remain normal and healthy.

Gene Regulation

The genome of a given organism contains thousands of genes, but not all these genes need to be active at any given moment. A gene is "expressed" when it is being transcribed into mRNA (and translated into protein), and there exist many cellular methods of controlling the expression of genes such that proteins are produced only when needed by the cell. Transcription factors are regulatory proteins that bind to the start of genes, either promoting or inhibiting the transcription of the gene. Within the genome of *Escherichia coli* bacteria, for example, there exists a series of genes necessary for the synthesis of the amino acid tryptophan. However, when tryptophan is already available to the cell, these genes for tryptophan synthesis are no longer needed. The presence of tryptophan directly affects the activity of the genes — the tryptophan molecules bind to the tryptophan repressor (a transcription factor), changing the repressor's structure such that the repressor is "active" and binding to the genes. The tryptophan repressor blocks the transcription and prevents expression of the genes, thus creating negative feedback regulation of the tryptophan synthesis process.

Differences in gene expression are especially clear within multicellular organisms, where cells all contain the same genome but have very different structures and behaviors due to the expression of different sets of genes. All the cells in a multicellular organism derive from a single cell, differentiating into different cell types in response to external and intercellular signals and gradually establishing different patterns of gene expression to create different behaviors. No single gene is responsible for the development of structures within multicellular organisms, these patterns arise from the complex interactions between many cells.

Within eukaryotes there exist structural features of chromatin which influence the transcription of genes, often in the form of modifications to DNA and chromatin that are stably inherited by daughter cells. These features are called "epigenetic" because they exist "on top" of the DNA sequence and retain inheritance from one cell generation to the next. Because of epigenetic features, different cell types grown within the same medium can retain very different properties. Although epigenetic features are generally dynamic over the development of organisms, some, like the phenomenon of paramutation, have multigenerational inheritance and exist as rare exceptions to the general rule of DNA as the basis for inheritance.

Mutations and Natural Selection

Mutations

During the process of DNA replication, errors occasionally occur in the polymerization of the second strand (these error rates are generally extremely low, 1 error in every 10-100 million bases). These errors, called mutations, can have an impact on the phenotype of an organism, especially if they occur within the protein coding sequence of a gene. Processes which increase the rate of changes in DNA are called "mutagenic": mutagenic chemicals promote errors in DNA replication, often by interfering with the structure of base-pairing, while UV radiation induces mutations by causing damage to the DNA structure. Chemical damage to DNA occurs naturally as well, and cells use DNA repair mechanisms to repair mismatches and breaks in DNA — nevertheless, the repair sometimes fails to return the DNA to its original sequence.

In organisms which use chromosomal crossover to exchange DNA and shuffle genes, errors in alignment during meiosis can also cause mutations. Errors in crossover are especially likely when similar sequences cause partner chromosomes to adopt a mistaken alignment, which makes some regions in genomes more prone to mutating in this way. These errors create large structural changes in DNA sequence — duplications, inversions or deletions of entire regions, or the accidental exchanging of whole parts between different chromosomes (called "translocation").

Natural Selection and Evolution

Mutations produce organisms with different genotypes, and those differences can result in different phenotypes. Many genetic mutations, called "neutral mutations", have a negligible effect on an organism's phenotype, health, and reproductive fitness. Mutations which do have an effect are often deleterious, but occasionally mutations arise which are beneficial in the current environmental context of the organism.

Population genetics research studies the distributions of these genetic differences within populations and how the distributions change over time. Changes in the frequency of an allele in a population can be influenced by natural selection, where a given allele's higher rate of survival and reproduction causes it to become more frequent in the population over time. Genetic drift can also occur, where chance events lead to random changes in allele frequency.

Over many generations, the genomes of organisms can change, resulting in the phenomenon of evolution. Mutations and the selection for beneficial mutations can cause a species to evolve into forms that better survive their environment, a process called adaptation. New species are formed through the process of speciation, a process often caused by geographical separations that allow different populations to genetically diverge.

As sequences diverge and change during the process of evolution, these differences between sequences can be used as a molecular clock to calculate the evolutionary distance between them. Genetic comparisons are generally considered the most accurate method of characterizing the relatedness between species, an improvement over the sometimes deceptive comparison of phenotypic characteristics. The evolutionary distances between species can be combined to form evolutionary trees. These trees are commonly considered the most accurate representation of relatedness, although the transfer of genetic material between unrelated species (known as "horizontal gene transfer" and most common in bacteria) cannot be represented by the tree.

Genetic Research and Technology

Model Organisms and Genetics

Although geneticists originally studied inheritance in a wide range of organisms, researchers began to specialize in studying the genetics of a particular subset of organisms. The fact that significant research already existed for a given organism would encourage new researchers to choose it for further study, and so eventually a few "model organisms" became the basis for most genetics research. Common research topics in model organism genetics include the study of gene regulation and the involvement of genes in development and cancer.

Organisms were chosen, in part, for convenience — short generation times and facile genetic manipulation made some organisms popular genetics research tools. Widely used model organisms include the gut bacterium *Escherichia coli*, the plant *Arabidopsis thaliana*, baker's yeast (*Saccharomyces cerevisiae*), the nematode *Caenorhabditis elegans*, the common fruit fly (*Drosophila melanogaster*), and the common house mouse (*Mus musculus*).

Medical Genetics Research

Medical genetics research seeks to find and study the genetic causes of human diseases. When searching for an unknown gene that may be involved in a disease, researchers commonly use genetics linkage and genetic pedigree charts to find the location on the genome associated with the disease. At the population level, researchers take advantage of Mendelian randomization to look for locations in the genome that are associated with diseases, a technique especially useful for multigenic traits not clearly defined by a single gene. Once a candidate gene is found, further research is often done on the same gene (called an orthologous gene) in model organisms.

Although many inherited diseases are subjects of genetic research, cancer is also considered a genetic disease. The process of cancer development in the body is a combination of events. Mutations occasionally occur within cells in the body as they divide — while these mutations will not be inherited by any children, they can effect the behavior of cells, sometimes causing them to grow and divide more frequently. There are biological mechanisms that attempt to stop this process — signals are given to inappropriately dividing cells that should trigger cell death, but sometimes additional mutations occur that cause cells to ignore these messages. An internal process of natural selection occurs within the body and eventually mutations accumulate within cells to promote their own growth, creating a cancerous tumor that grows and invades various tissues of the body.

Genetic Technologies

A variety of techniques exist for manipulating DNA in the laboratory. Restriction enzymes are a commonly used enzyme that cuts DNA at specific sequences, producing predictable fragments of DNA. The use of ligation enzymes allows these fragments to be stitched back together, and by ligating fragments of DNA together from different sources, researchers can create recombinant DNA. Often associated with genetically modified organisms, recombinant DNA is commonly used in the context of plasmids — short circular DNA fragments with a few genes on them. By inserting plasmids into bacteria and growing those bacteria on plates of agar (to isolate clones of bacteria cells), researchers can clonally amplify the inserted fragment of DNA (a process known as molecular cloning). (Cloning can also refer to the creation of clonal organisms, through various techniques.)

DNA can also be amplified using a procedure called the polymerase chain reaction (PCR). By using specific short sequences of DNA, PCR can exponentially amplify a targeted region of DNA, isolating and amplifying a selected section of DNA. Because it can amplify from extremely small amounts of DNA, PCR is also often used to detect the presence of specific DNA sequences.

DNA Sequencing & Genomics

One of the most fundamental technologies developed to study genetics, DNA sequencing allows researchers to determine the sequence of nucleotides in DNA fragments. Developed in 1977 by Frederick Sanger and coworkers, chain-termination sequencing is now routinely used to sequence DNA fragments. With this technology, researchers have been able to study the molecular sequences associated with many human diseases. As sequencing has become less expensive and with the aid of computational tools, researchers have sequenced the genomes of many organisms by stitching together the sequences of many different fragments (a process called "genome assembly"). These technologies were used to sequence the human genome, leading to the completion of the Human genome project in 2003.

The large amount of sequences available has created the field of "genomics," research which uses computational tools to search for and analyze patterns in the full genomes of organisms. Genomics can also be considered a subfield of bioinformatics, which uses computational approaches to analyze of large sets of biological data.

GENETIC DISORDER

A genetic disorder is a condition caused by abnormalities in genes or chromosomes. While some diseases, such as cancer, are ndue to genetic abnormalities acquired in a few cells during life, the term "genetic disease" most commonly refers to diseases present in all cells of the body and present since conception. Some genetic disorders are caused by chromosomal abnormalities due to errors in meiosis, the process which produces reproductive cells such as sperm and eggs. Examples include Down syndrome (extra chromosome 21), Turner Syndrome (45X0) and Klinefelter's syndrome (a male with 2 X chromosomes). Other genetic changes may occur during the production of germ cells by the parent. One example is the triplet expansion repeat mutations which can cause fragile X syndrome or Huntington's disease. Defective genes may also be inherited intact from the parents. In this case, the genetic disorder is known as a hereditary disease. This can often happen unexpectedly when two healthy carriers of a defective recessive gene reproduce, but can also happen when the defective gene is dominant.

Currently about 4,000 genetic disorders are known, with more being discovered. Most disorders are quite rare and affect one person in every several thousands or millions. Cystic fibrosis is one of the most common genetic disorders; around 5% of the population of the United States carry at least one copy of the defective gene. Some types of recessive gene disorder confer an advantage in the heterozygous state in certain environments.

Genetic diseases are typically diagnosed and treated by geneticists. Genetic counselors assist the physicians and directly counsel patients. The study of genetic diseases is a scientific discipline whose theoretical underpinning is based on population genetics.

Single Gene Disorders

Where genetic disorders are the result of a single mutated gene they can be passed on to subsequent generations in the ways outlined in the table below. Genomic imprinting and uniparental disomy, however, may affect inheritance patterns. The divisions between recessive and dominant are not "hard and fast" although the divisions between autosomal and X-linked are (related to the position of the gene). For example, achondroplasia is typically considered a dominant disorder, but children with two genes for achondroplasia have a severe skeletal disorder that achondroplasics could be viewed as carriers of. Sickle-cell anemia is also considered a recessive condition, but carriers that have it by half along with the normal gene have increased immunity to malaria in early childhood, which could be described as a related dominant condition.

Multifactorial and Polygenic Disorders

Genetic disorders may also be complex, multifactorial or polygenic, this means that they are likely associated with the effects of multiple genes in combination with lifestyle and environmental factors. Multifactoral disorders include heart disease and diabetes. Although complex disorders often cluster in families, they do not have a clear-cut pattern of inheritance. This makes it difficult to determine a person's risk of inheriting or passing on these disorders. Complex disorders are also difficult to study and treat because the specific factors that cause most of these disorders have not yet been identified.

On a pedigree, polygenic diseases do tend to "run in families", but the inheritance does not fit simple patterns as with Mendelian diseases. But this does not mean that the genes cannot eventually be located and studied. There is also a strong environmental component to many of them (e.g., blood pressure).

- autism
- heart disease
- hypertension
- diabetes
- obesity
- cancers
- cleft palate
- Mental retardation

HISTORY OF GENETICS

The history of genetics is generally held to have started with the work of an Augustinian monk, Gregor Mendel. His work on pea plants, published in 1866, described what came to be known as Mendelian inheritance. In the centuries before—and for several decades after—Mendel's work, a wide variety of theories of heredity proliferated. 1900 marked the "rediscovery of Mendel" by Hugo de Vries, Carl Correns and Erich von Tschermak, and by 1915 the basic principles of Mendelian genetics had been applied to a wide variety of organisms—most notably the fruit fly *Drosophila melanogaster*. Led by Thomas Hunt Morgan and his fellow "drosophilists", geneticists developed the Mendelian-chromosome theory of heredity, which was widely accepted by 1925. Alongside experimental work, mathematicians developed the statistical framework of population genetics, bring genetical explanations into the study of evolution.

With the basic patterns of genetic inheritance established, many biologists turned to investigations of the physical nature of the gene. In the 1940s and early 1950s, experiments pointed to DNA as the portion of chromosomes (and perhaps other nucleoproteins) that held genes. A focus on new model organisms such as viruses and bacteria, along with the discovery of the double helical structure of DNA in 1953, marked the transition to the era of molecular genetics. In the following years, chemists developed techniques for sequencing both nucleic acids and proteins, while others worked out the relationship between the two forms of biological molecules: the genetic code. The regulation of gene expression became a central issue in the 1960s; by the 1970s gene expression could be controlled and manipulated through genetic engineering. In the last decades of the 20th century, many biologists focused on large-scale genetics projects, sequencing entire genomes.

Pre-Mendelian Ideas on Heredity

Ancient Theories

The most influential early theories of heredity were that of Hippocrates and Aristotle. Hippocrates' theory (possibly based on the teachings of Anaxagoras) was similar to Darwin's later ideas on pangenesis, involving heredity material that collects from throughout the body. Aristotle suggested instead that the (nonphysical) form-giving principle of an organism was transmitted through semen, determining the shape of the female's menstrual blood through an organism's early development. For both Hippocrates and Aristotle—and nearly all Western scholars through to the late 19th century—the inheritance of acquired characters was a supposedly well-established fact that any adequate theory of heredity had to explain. At the same time, individual species were taken to have a fixed essence; such inherited changes were merely superficial.

Plant Systematics and Hybridization

In the 18th century, with increased knowledge of plant and animal diversity and the accompanying increased focus on taxonomy, new ideas about heredity began to appear. Linnaeus and others (among them Joseph Gottlieb Kölreuter, Carl Friedrich von Gärtner, and Charles Naudin) conducted extensive experiments with hybridization, especially species hybrids. Species hybridizers described a wide variety of inheritance phenomena, include hybrid sterility and the high variability of back-crosses.

Plant breeders were also developing an array of stable varieties in many important plant species. In the early 19th century, Augustin Sageret established the concept of dominance, recognizing that when some plant varieties are crossed, certain characters (present in one parent) usually appear in the offspring; he also found that some ancestral characters found in neither parent may appear in offspring. However, plant breeders made little attempt to develop a theoretical foundation for their work or to integrate their knowledge with work in physiology and natural history.

Mendel

In breeding experiments between 1856 and 1865, Gregor Mendel first traced inheritance patterns of certain traits in pea plants and showed that they obeyed simple statistical rules. Although not all features show these patterns of Mendelian inheritance, his work acted as a proof that application of statistics to inheritance could be highly useful. Since that time many more complex forms of inheritance have been demonstrated.

From his statistical analysis Mendel defined a concept that he described as an *allele*, which was the fundamental unit of heredity. The term *allele* as Mendel used it is nearly synonymous with the term *gene*, and now means a specific variant of a particular gene.

Mendel's work was published in 1866 as *Experiments on Plant Hybridization* in the *Proceedings of the Natural History Society of Brünn*, following two lectures he gave on the work in early 1865.

Post-Mendel, Pre-re-Discovery

Mendel's work was published in a relatively obscure scientific journal, and it was not given any attention in the scientific community. Instead, discussions about modes of heredity were galvanized by Darwin's theory of evolution by natural selection, in which mechanisms of non-Lamarckian heredity seemed to be required. Darwin's own theory of heredity, pangenesis, did not meet with any large degree of acceptance. A more mathematical version of pangenesis, one which dropped much of Darwin's Lamarckian holdovers, was developed as the "biometrical" school of heredity by Darwin's cousin, Francis Galton. Under Galton and his successor Karl Pearson, the biometrical school attempted to build statistical models for heredity and evolution, with some limited but real success, though the exact methods of heredity were unknown and largely unquestioned.

Classical Genetics

The significance of Mendel's work was not understood until early in the twentieth century, after his death, when his research was re-discovered by other scientists working on similar problems. Hugo de Vries, Carl Correns and Erich von Tschermak

There was then a feud between Bateson and Pearson over the hereditary mechanism. Fisher solved this in The Correlation Between Relatives on the Supposition of Mendelian Inheritance

1865 Gregor Mendel's paper, *Experiments on Plant Hybridization*

1869 Friedrich Miescher discovers a weak acid in the nuclei of white blood cells that today we call DNA

1880-1890 Walther Flemming, Eduard Strasburger, and Edouard van Beneden elucidate chromosome distribution during cell division

1889 Hugo de Vries postulates that "inheritance of specific traits in organisms comes in particles", naming such particles "(pan)genes"

1903 Walter Sutton hypothesizes that chromosomes, which segregate in a Mendelian fashion, are hereditary units

1905 William Bateson coins the term "genetics" in a letter to Adam Sedgwick and at a meeting in 1906

1908 Hardy-Weinberg law derived.

1910 Thomas Hunt Morgan shows that genes reside on chromosomes

1913 Alfred Sturtevant makes the first genetic map of a chromosome

1913 Gene maps show chromosomes containing linear arranged genes

1918 Ronald Fisher publishes "The Correlation Between Relatives on the Supposition of Mendelian Inheritance" the modern synthesis of genetics and evolutionary biology starts. See population genetics.

1928 Frederick Griffith discovers that hereditary material from dead bacteria can be incorporated into live bacteria

1931 Crossing over is identified as the cause of recombination

1933 Jean Brachet is able to show that DNA is found in chromosomes and that RNA is present in the cytoplasm of all cells.

1941 Edward Lawrie Tatum and George Wells Beadle show that genes code for proteins; see the original central dogma of genetics

The DNA Era

1944 Oswald Theodore Avery, Colin McLeod and Maclyn McCarty isolate DNA as the genetic material (at that time called transforming principle)

1950 Erwin Chargaff shows that the four nucleotides are not present in nucleic acids in stable proportions, but that some general rules appear to hold (e.g., that the amount of adenine, A, tends to be equal to that of thymine, T). Barbara McClintock discovers transposons in maize

1952 The Hershey-Chase experiment proves the genetic information of phages (and all other organisms) to be DNA

1953 DNA structure is resolved to be a double helix by James D. Watson and Francis Crick

1956 Joe Hin Tjio and Albert Levan established the correct chromosome number in humans to be 46

1958 The Meselson-Stahl experiment demonstrates that DNA is semiconservatively replicated

1961-1967 Combined efforts of scientists "crack" the genetic code, including Marshall Nirenberg, Har Gobind Khorana, Sydney Brenner & Francis Crick

1964 Howard Temin showed using RNA viruses that the direction of DNA to RNA transcription can be reversed

1970 Restriction enzymes were discovered in studies of a bacterium, *Haemophilus influenzae*, enabling scientists to cut and paste DNA

The Genomics Era

1972, Walter Fiers and his team at the Laboratory of Molecular Biology of the University of Ghent (Ghent, Belgium) were the first to determine the sequence of a gene: the gene for Bacteriophage MS2 coat protein.

1976, Walter Fiers and his team determine the complete nucleotide-sequence of bacteriophage MS2-RNA

1977 DNA is sequenced for the first time by Fred Sanger, Walter Gilbert, and Allan Maxam working independently. Sanger's lab sequence the entire genome of Bacteriophage Ö-X174.

1983 Kary Banks Mullis discovers the polymerase chain reaction enabling the easy amplification of DNA

1989 The human gene that encodes the CFTR protein was sequenced by Francis Collins and Lap-Chee Tsui. Defects in this gene cause cystic fibrosis.

1995 The genome of *Haemophilus influenzae* is the first genome of a free living organism to be sequenced

1996 *Saccharomyces cerevisiae* is the first eukaryote genome sequence to be released

1998 The first genome sequence for a multicellular eukaryote, *Caenorhabditis elegans*, is released

2001 First draft sequences of the human genome are released simultaneously by the Human Genome Project and Celera Genomics.

2003 (14 April) Successful completion of Human Genome Project with 99% of the genome sequenced to a 99.99% accuracy

MENDELIAN INHERITANCE

Mendelian inheritance (or Mendelian genetics or Mendelism) is a set of primary tenets relating to the transmission of hereditary characteristics from parent organisms to their children; it underlies much of genetics. They were initially derived from the work of Gregor Mendel published in 1865 and 1866 which was "re-discovered" in 1900, and were initially very controversial. When they were integrated with the chromosome theory of inheritance by Thomas Hunt Morgan in 1915, they became the core of classical genetics.

History

The laws of inheritance were derived by Gregor Mendel, a 19th century Moravian monk conducting plant hybridity experiments. Between 1856 and 1863, he cultivated and tested some 28,000 pea plants. His experiments brought forth two generalizations which later became known as *Mendel's Laws of Heredity* or *Mendelian inheritance.* These are described in his essay "Experiments on Plant Hybridization" that was read to the Natural History Society of Brno on February 8 and March 8, 1865, and was published in 1866.

Mendel's results were largely rejected. Though they were not completely unknown to biologists of the time, they were not seen as being crucial. Even Mendel himself did not see their ultimate applicability, and thought they only applied to certain categories of species. In 1900, however, the work was "re-discovered" by three European scientists, Hugo de Vries, Carl Correns, and Erich von Tschermak. The exact nature of the "re-discovery" has been somewhat debated: De Vries published first on the subject, and Correns pointed out Mendel's priority after having read De Vries's paper and realizing that he himself did not have priority, and De Vries may not have acknowledged truthfully how much of his knowledge of the laws came from his own work, or came only after reading Mendel's paper. Later scholars have accused Von Tschermak of not truly understanding the results at all.

Regardless, the "re-discovery" made Mendelism an important but controversial theory. Its most vigorous promoter in Europe was William Bateson, who coined the term "genetics", "gene", and "allele" to describe many of its tenets. The model of heredity was highly contested by other biologists because it implied that heredity was discontinuous, in opposition to the apparently continuous variation observable. Many biologists also dismissed the theory because they were not sure it would apply to all species, and there seemed to be very few true Mendelian characters in nature. However later work by biologists and statisticians such as R.A. Fisher showed that if multiple Mendelian factors were involved for individual traits, they could produce the diverse amount of results observed in nature. Thomas Hunt Morgan and his assistants would later integrate the theoretical model of Mendel with the chromosome theory of inheritance, in which the chromosomes of cells were thought to hold the actual hereditary particles, and create what is now known as classical genetics, which was extremely successful and cemented Mendel's place in history.

Mendel's findings allowed other scientists to simplify the emergence of traits to mathematical probability. A large portion of Mendel's findings can be traced to his choice to start his experiments only with true breeding plants. He also only measured absolute characteristics such as color, shape, and position of the offspring. His data was expressed numerically and subjected to statistical analysis. This method of data reporting and the large sampling size he used gave credibility to his data. He also had the foresight to look through several successive generations of his pea plants and record their variations. Without his careful attention to procedure and detail, Mendel's work could not have had the impact it made on the world of genetics.

Mendel's Laws

Law of Segregation

The Law of Segregation, also known as Mendel's First Law, essentially has four parts.

1. Alternative versions of genes account for variations in inherited characteristics. This is the concept of alleles. Alleles are different versions of genes that impart the same characteristic. For example, each human has a gene that controls eye color, but there are variations among these genes in accordance with the specific color for which the gene "codes".

2. For each characteristic, an organism inherits two alleles, one from each parent. This means that when somatic cells are produced from two alleles, one allele comes from the mother and one from the father. These alleles may be the same.

3. If the two alleles differ, then one, the allele that encodes the dominant trait, is fully expressed in the organism's appearance; the other, the allele encoding the recessive trait, has no noticeable effect on the organism's appearance. In other words, only the dominant trait is seen in the phenotype of the organism. This allows recessive traits to be passed on to offspring even if they are not expressed. Not all traits have a dominant-recessive relationship, however. The petal color of flowers of the "Japanese four o'clock" plant *Mirabilis jalapa* illustrate incomplete dominance. There is also codominance e.g. human blood types, for which A and B are codominant and O is recessive.

4. The two alleles for each characteristic segregate during gamete production. This means that each gamete will contain only one allele for each gene. This allows the maternal and paternal alleles to be combined in the offspring, ensuring variation.

It is often misconstrued that the gene itself is dominant, recessive, codominant, or incompletely dominant. It is, however the *trait*, or gene product that the allele encodes that is dominant, etc.

Law of Independent Assortment

The Law of Independent Assortment, also known as "Inheritance Law" or Mendel's Second Law, states that the inheritance pattern of one trait will not affect the inheritance pattern of another. While his experiments with mixing one trait always resulted in a 3:1 ratio between dominant and recessive phenotypes, his experiments with mixing two traits (dihybrid cross) showed 9:3:3:1 ratios. But the shows that each of the two genes are independently inherited with a 3:1 ratio. Mendel concluded that different traits are inherited independently of each other, so that there is no relation, for example, between a cat's color and tail length. This is actually only true for genes that are not linked to each other.

Background

The reason for these laws is found in the nature of the cell nucleus. It is made up of several chromosomes carrying the genetic traits. In a normal cell, each of these chromosomes has two parts, the chromatids. A reproductive cell, which is created in a process called meiosis, usually contains only one of those chromatids of each chromosome. By merging two of these cells (usually one male and one female), the full set is restored and the genes are mixed. The resulting cell becomes a new embryo. The fact that this new life has half the genes of each parent (23 from mother, 23 from father for total of 46) is one reason for the Mendelian laws. The second most important reason is the varying dominance of different genes, causing some traits to appear unevenly instead of averaging out (whereby dominant doesn't mean more likely to reproduce - recessive genes can become the most common, too).

There are several advantages of this method (sexual reproduction) over reproduction without genetic exchange:

1. Instead of nearly identical copies of an organism, a broad range of offspring develops, allowing more different abilities and evolutionary strategies.
2. There are usually some errors in every cell nucleus. Copying the genes usually adds more of them. By distributing them randomly over different chromosomes and mixing the genes, such errors will be distributed unevenly

over the different children. Some of them will therefore have only very few such problems. This helps reduce problems with copying errors somewhat.

3. Genes can spread faster from one part of a population to another. This is for instance useful if there's a temporary isolation of two groups. New genes developing in each of the populations don't get reduced to half when one side replaces the other, they mix and form a population with the advantages of both sides.
4. Sometimes, a mutation (e. g. sickle cell anemia) can have positive side effects (in this case malaria resistance). The mechanism behind the Mendelian laws can make it possible for some offspring to carry the advantages without the disadvantages until further mutations solve the problems.

Mendelian Trait

A Mendelian trait is one that is controlled by a single locus and shows a simple Mendelian inheritance pattern. In such cases, a mutation in a single gene can cause a disease that is inherited according to Mendel's laws. Examples include sickle-cell anemia, Tay-Sachs disease, cystic fibrosis and xeroderma pigmentosa. A disease controlled by a single gene contrasts with a multi-factorial disease, like arthritis, which is affected by several loci (and the environment) as well as those diseases inherited in a non-Mendelian fashion. The Mendelian Inheritance in Man database is a catalog of, among other things, genes in which Mendelian mutants causes disease.

LIST OF GENETICISTS

This is a list of people who have made notable contributions to genetics. The growth and development of genetics represents the work of many people. This list of geneticists is therefore by no means complete. Contributors of great distinction to genetics are not yet on the list.

- Dagfinn Aarskog (1928-), Norwegian pediatrician and geneticist, described Aarskog-Scott syndrome
- Jon Aase (1936-), US dysmorphologist, described Aase syndrome, expert on fetal alcohol syndrome
- John Abelson (c.1939-), US biochemist, studies of machinery and mechanism of RNA splicing
- Susan L. Ackerman, US neurogeneticist, genes controlling brain development and neuron survival

- Jerry Adams (1940-), US molecular biologist in Australia, hematopoietic genetics and cancer
- Bruce Alberts (1938-), US biochemist, phage worker, studied DNA replication and cell division
- William Allan (1881-1943), US country doctor, pioneered human genetics
- C. David Allis (1951-), US biologist with a fascination for chromatin
- Carl-Henry Alström (1907-1993), Swedish psychiatrist, described genetic disease: Alstrom syndrome
- Sidney Altman (1939-), Canadian-US biophysicist who won Nobel Prize for catalytic functions of RNA
- Cecil A. Alport (1880-1959),UK internist, identified Alport syndrome (hereditary nephritis and deafness)
- David Altshuler (c.1965-), US endocrinologist and geneticist, the genetics of type 2 diabetes
- Bruce Ames (1928-), US molecular geneticist, created Ames test to screen chemicals for mutagenicity
- D. Bernard Amos (1923-2003), UK-US immunologist who studied the genetics of individuality
- Edgar Anderson (1897-1969), eminent US botanical geneticist
- E.G. ("Andy") Anderson, US Drosophila and maize geneticist
- William French Anderson (1936-), US worker in gene therapy
- Corino Andrade (1906-2005), Portuguese neurologist and clinical geneticist
- Tim Anson (1901-1968), US molecular biologist, proposed protein folding a reversible two-state reaction
- Stylianos E. Antonarakis (1951-), US-Greek medical geneticist, genotypic and phenotypic variation
- Werner Arber (1929-), Swiss microbiologist, Nobel Prize for discovery of restriction endonucleases
- Michael Ashburner (1942-), British Drosophila geneticist and polymath
- William Astbury (1898-1961), UK molecular biologist, X-ray crystallography of proteins and DNA
- Giuseppe Attardi, Italian-US molecular biologist, genetics of human mitochondrial function
- Charlotte Auerbach (1899-1994), German-born British pioneer in mutagenesis

- Oswald Avery (1877–1955), Canadian-born US co-discoverer that DNA is the genetic material
- Richard Axel (1946-) US physician-scientist, Nobel Prize for genetic analysis of olfactory system
- E. B. Babcock (1877-1954), US plant geneticist, pioneered genetic analysis of genus Crepis
- E-G Balbiani (1823-1899), French embryologist who found chromosome puffs now called Balbiani rings
- David Baltimore (1938-), US biologist, Nobel Prize for the discovery of reverse transcriptase
- Guido Barbujani (1955-), Italian population geneticist and evolutionary biologist
- Murray Barr (1908–1995), Canadian scientist, first saw Barr body in cells due to inactive X chromosome
- Cornelia Bargmann, US, molecular neurogeneticist studying the C. elegans brain
- David P. Bartel (B.A. 1982), US geneticist, discovered many microRNAs regulating gene expression
- William Bateson (1861-1926), British geneticist who coined the term "genetics"
- E. Baur (1875-1933), German geneticist, botanist, discovered inheritance of plasmids
- George Beadle (1903-1989), US Neurospora geneticist and Nobel Prize-winner
- Peter Emil Becker (1908-2000), German human geneticist, described Becker's muscular dystrophy
- Jon Beckwith, US microbiologist and geneticist, isolated first gene from a bacterial chromosome
- Peter Beighton (1934-) UK/South Africa medical geneticist, first warned of economy class syndrome
- Julia Bell (1879-1979), English geneticist who documented inheritance of many diseases
- John Belling (1866-1933), English cytogeneticist who developed staining technique for chromosomes
- Baruj Benacerraf (1920-), Venezuelan-US immunologist who won Nobel Prize for HLA system

- Kurt Benirschke (1924-), German-US pathologist, comparative cytogenetics, twinning in armadillos
- Seymour Benzer (1921-), US molecular biologist and pioneer of neurogenetics
- Paul Berg (1926-), US biochemist and Nobel Prize-winner for basic research on nucleic acids
- J.D. Bernal (1901-1971), Irish physicist and pioneer X-ray crystallographer
- James Birchler, Drosophila and Maize geneticists and cytogenticist.
- J. Michael Bishop (1936-), US microbial immunogeneticist, Nobel Prize-winner for oncogenes
- Elizabeth Blackburn (1948-), Australo-US biologist, Lasker Award on telomeres and telomerase
- Günter Blobel (1936-), German-US biologist, Nobel Prize for protein targeting (address tags on proteins)
- David Blow (1931-2004), British biophysicist who helped develop X-ray crystallography of proteins
- Baruch Blumberg (Barry Blumberg) (1925-), US physician and Nobel Prize-winner on hepatitis B
- Julia Bodmer (1934-2001), British geneticist, key figure in discovery and definition of the HLA system
- Walter Bodmer (1936-), German-UK human population geneticist, immunogeneticist, cancer research
- James Bonner (1910-1996), far-ranging US molecular biologist, into histones, chromatin, nucleic acids
- David Botstein (1942-), Swiss-born US molecular geneticist, brother of Leon Botstein
- Theodor Boveri (1862-1915), German biologist and cytogeneticist
- Peter Bowen (1932-1988), Canadian medical geneticist
- Herb Boyer (1936-), US, created transgenic bacteria inserting human insulin gene into E. coli
- Paul D. Boyer (1918-), US biochemist and Nobel Prize-winner
- Jean Brachet (1909-1998), Belgian biochemist, made key contributions to fathoming roles of RNA
- Roscoe Brady US physician-scientist at NIH, studies of genetic neurological metabolic disorders

- Sydney Brenner (1927-), British molecular biologist and Nobel Prize-winner
- Calvin Bridges (1889-1938), US geneticist, non-disjunction proof that chromosomes contain genes
- R.A. Brink (1897-1984), Canadian-US plant geneticist and breeder, studied paramutation, transposons
- Roy Britten (1919-) US molecular and evolutionary biologist, discovered and studied junk DNA
- John Brookfield Drosophila population geneticist.
- Michael Stuart Brown (1941-) US geneticist and Nobel Prize-winner on cholesterol metabolism
- Manuel Buchwald (1940-), Peruvian-born Canadian medical geneticist and molecular geneticist
- Linda Buck (1947-) US biologist, Nobel Prize for post-doc work (with Axel) cloning olfactory receptors
- James Bull, US molecular biologist and phage worker, evolution of sex determining mechanisms
- Luther Burbank (1849-1926), US botanist, horticulturist, pioneer in agricultural science
- Macfarlane Burnet (1899-1985), Australian biologist, Nobel Prize for immunological tolerance
- Cyril Burt (1883-1971), British educational psychologist, did debated mental and behavioral twin study
- John Cairns (1922-), UK physician-scientist, showed bacterial DNA one molecule with replicating fork
- Allan Campbell, US microbiologist and geneticist, pioneering work on phage lambda
- Howard Cann, US pediatrician and geneticist, human population genetics at Stanford and CEPH in Paris
- Antonio Cao (1929-), Italian pediatrician and medical geneticist, expert on the thalassemias
- Mario Capecchi (1937-), Italian-born US molecular geneticist, co-invented the knock-out mouse, Nobel prize for Medicine, 2007
- Elof Axel Carlson, US geneticist and eminent historian of science

- Gyaneshwer Chaubey, Indian population geneticist.
- Hampton Carson (1914-2004), US population geneticist, studied cytogenetics and evolution of Drosophila
- Tom Caskey (c.1938-), US internist, human geneticist and entrepreneur; biochemical diseases
- Torbjörn Caspersson (1910-1997), Swedish cytogeneticist, revealed human chromosome banding
- William B. Castle (1897-1990), US hematologist, work on hereditary spherocytosis, sickle cell anemia
- William E. Castle (1867-1962), US geneticist, inspired T.H. Morgan, father of William B. Castle
- David Catcheside (1907-1994) UK plant geneticist, expert on genetic recombination, active in Australia
- Bruce Cattanach (1932-), eminent UK mouse geneticist, X-inactivation and sex determination in mice
- Luigi Luca Cavalli-Sforza (1922-), distinguished Italian population geneticist at Stanford University
- Thomas Cech (1947-), US biochemist who won Nobel Prize for catalytic functions of RNA
- Aravinda Chakravarti (1954-), Indian-born bioinformatician studying genetic factors in common diseases
- Jean-Pierre Changeux (1936-), French molecular neurobiologist, studied allosteric proteins
- Erwin Chargaff (1905-2002), Austrian-born US biochemist, Chargaff's rules led to the double helix
- Brian Charlesworth (1945-), British evolutionary biologist, husband of Deborah Charlesworth
- Deborah Charlesworth, British evolutionary biologist, wife of Brian Charlesworth
- Martha Chase (1927-2003), US biologist, with Hersey proved genetic material is DNA, not protein
- Sergei Chetverikov (1880-1959), Russian population geneticist
- Barton Childs (1916-), US pediatrician, biochemical geneticist, philosopher of medical genetics

- George Church (1954-), US molecular geneticist, did first direct genomic sequencing with Gilbert
- Aaron Ciechanover (1947-), Israeli biologist, won Nobel Prize for ubiquitin-mediated protein degradation
- Bryan Clarke (1932-), British population geneticist, studied apostatic selection and molecular evolution
- Cyril Clarke (1907-2000), British medical geneticist, discovered how to prevent Rh disease in newborns
- Jens Clausen (1891-1969), Danish-US botanist, geneticist, and ecologist
- Edward H. Coe, Jr. (1926-), influential US maize (corn) geneticist
- Stanley Cohen (1922-), US neurobiologist, Nobel Prize for cell growth factors
- Francis Collins (1950-), US medical geneticist, gene cloner, director of Human Genome Institute
- Robert Corey (1897-1971), US biochemist, á-helix, â-sheet and atomic models for proteins
- Carl Correns (1864-1933), German botanist and geneticist, one of the re-discoverers of Mendel in 1900
- Lewis L. Coriell (1911-2001), US pioneer in culturing human cells
- Diane W. Cox, Canadian medical geneticist and expert on Wilson disease
- Harriet Creighton (1909-2004), US botanist who with McClintock first saw chromosomal crossover
- Francis Crick (1916-2004), English molecular biologist, neuroscientist, co-discoverer of the double helix
- James F. Crow (1916-), US population geneticist and renowned teacher of genetics
- Lucien Cuenot (1886-1901), French biologist, proved Mendel's rules apply to animals as well as plants
- A. Jamie Cuticchia (1966-), US geneticist, into human genome informatics
- David M. Danks (1931-2003), Australian pediatrician and medical geneticist, expert on Menkes disease
- C. D. Darlington (1903-1981), British biologist and geneticist, elucidated chromosomal crossover
- Charles Darwin (1809-1882), English naturalist and author of *Origin of the Species*

- Kay Davies, English geneticist, expert on muscular dystrophy
- Jean Dausset (1916-) French immunogeneticist and Nobel Prize-winner for the HLA system
- Martin Dawson (1896-1945), Canadian-US researcher, confirmed and named genetic transformation
- Margaret Dayhoff (1925-1983), US pioneer in bioinformatics of protein sequences and evolution
- Albert de la Chapelle (1933-), eminent Finnish medical geneticist, genetic predisposition to cancer
- Max Delbruck (1906-1981), German-US scientist, Nobel Prize for genetic structure of viruses
- Charles DeLisi, US biophysicist, led the initiative that planned and launched the Human Genome Project
- Félix d'Herelle (1873-1949), Canadian-French microbiologist, discovered phages, invented phage therapy
- Hugo de Vries (1848-1935), Dutch botanist and one of the re-discoverers of Mendel's laws in 1900
- M. Demerec (1895-1966), Croatian-US geneticist, directed Cold Spring Harbor Laboratory
- Theodosius Dobzhansky (1900-1975), noted Ukrainian-US geneticist and evolutionary biologist
- John Doebley, US geneticist, studies genes that drive development and evolution of plants
- Peter Doherty (1940-), Australian, won Nobel Prize for immune recognition of antigens
- Albert Dorfman (1916-1982), US biochemical geneticist, discovered cause of Hurler's syndrome
- Gabriel Dover, British evolutionary geneticist
- NT Dubinin (1907-1998), Russian biologist and geneticist
- Bernard Dutrillaux (1940-), French cytogeneticist, chromosome banding, comparative cytogenetics
- Christian de Duve (1917-), Belgian cytologist, Nobel Prize for cell organelles (peroxisomes, lysosomes)
- A.W.F. Edwards (1935-), British statistician, geneticist, developed methods of phylogenetic analysis

- John Edwards (1928-), British medical geneticist and cytogeneticist who first described trisomy 18
- R. A. Emerson (1873–1947), American plant geneticist, the main pioneer of corn genetics
- Sterling Emerson (1900-1988), American, biochemical genetics, recombination, son of R. A. Emerson
- Alan Emery (1928-), British neuromuscular geneticist, Emery-Dreifuss muscular dystrophy
- Boris Ephrussi (1901–1979), Russian-born French geneticist, created way to transplant chromosomes
- Robert C. Elston (1932-), British-born American biostatistical genetics and genetic epidemiologist
- Charlie Epstein, American medical geneticist, mouse model, or, wounded by the Unabomber
- Herbert McLean Evans (1882-1971), US anatomist, reported in 1918 humans had 48 chromosomes
- Martin Evans, British scientist, discovered embryonic stem cells and developed knockout mouse
- Warren Ewens, Australian-US mathematical population geneticist, Ewens's sampling formula
- Alexander Cyril Fabergé (1912-1988), Russian-born Anglo-American geneticist, grandson of Carl Fabergé
- D. S. Falconer (1913-2004), Scottish quantitative geneticist, wrote textbook to the subject
- Stanley Falkow, US microbial geneticist, molecular mechanisms of bacterial pathogenesis
- Harold Falls (1909-2006), US ophthalmologic geneticist, helped found first genetics clinic in US
- William C. Farabee (1865-1925), US anthropologist, brachydactyly is evidence of Mendelism in humans
- Nina Fedoroff (c. 1945-), US plant geneticist, cloning of transposable elements, plant stress response
- Malcolm Ferguson-Smith (1931-) UK cytogeneticist, Klinefelter's syndrome, chromosome flow cytometry

- Philip J. Fialkow (1934-1996), US internist, educator, research in medical genetics and cancer genetics
- Giorgio Filippi (1935-1996), Italian medical geneticist, researched diseases linked to X chromosome
- J.R.S. Fincham (1926-2005), British microbial (Neurospora) and biochemical geneticist
- Gerald Fink (1941-), US molecular geneticist, preeminent figure in the field of yeast genetics
- Andrew Fire (1959-), US geneticist, Nobel Prize with Mello for discovery of RNA interference
- Robert L. Fischer (1950-), A US geneticist, contributed to the understanding of genomic imprinting and epigenetics
- R.A. Fisher (1890-1962), British stellar statistician, evolutionary biologist, and geneticist
- Ed Fischer (1920-), Swiss-US biochemist, Nobel Prize for phosphorylation as switch activating proteins
- Eugen Fischer (1874-1967), German physician, anthropologist, eugenicist, influenced Nazi racial hygiene
- Asbjorn Folling (1888-1973), Norwegian biochemist and physician who discovered phenylketonuria (PKU)
- E.B. Ford (1901-1988), British ecological geneticist, specializing in butterflies and moths
- Charles Ford (1912–1999), British pioneer in the golden age of mammalian cytogenetics
- Heinz Fraenkel-Conrat (1910-1999), German-born US biochemist who studied tobacco mosaic virus
- Rosalind Franklin (1920-1958), British crystallographer whose data led to discovery of double helix
- Clarke Fraser (1920-), Canada's first medical geneticist, student of congenital malformations
- Elaine Fuchs (c.1951-), US cell biologist, molecular mechanisms of skin diseases, reverse genetics
- Walter Fuhrmann (1924-1995), German medical geneticist, at Giessen University
- Douglas J. Futuyma (1942-), US evolutionary and ecological biologist

- Fred Gage, US neuroscientist, studies of neurogenesis and neuroplasticity of the adult brain
- Joseph G. Gall (1932-), distinguished US cell biologist, chromosomes, created in situ hybridization
- Francis Galton (1822-1911), British geneticist, eugenicist, statistician
- George Gamow (1904-1968), Ukrainian-born American polymath, proposed genetic code concept
- Eldon J. Gardner (1909-1989), US professor of genetics in Utah, described Gardner's syndrome
- Alan Garen (c.1924-), US, early molecular geneticist, nonsense triplets terminating transcription
- Archibald Garrod (1857-1936), English physician, pioneered inborn errors, founded biochemical genetics
- Stan Gartler (1923-), US human geneticist, G6PD as X-linked marker, HeLa cells contaminating cell lines
- Luigi Gedda (1902-2000), Italian geneticist best known for his fascination with twin studies
- Walter Gehring (1939-), Swiss, developmental genetics of Drosophila, discovered homeobox
- Park S. Gerald (1921-1993), US medical geneticist, research on hemoglobins and chromosomes
- James L. German, US medical geneticist & cytogeneticist, pioneer on Bloom syndrome
- Walter Gilbert (1932-), US biochemist and molecular biologist, Nobel Prize-winner, entrepreneur
- H. Bentley Glass (1906-2005) US geneticist, provocative science theorizer, writer, science policy maker
- Salome Gluecksohn-Waelsch (1907-), German-born US co-founder of developmental genetics
- Richard Goldschmidt (1878-1958),German-American, integrated genetics, development, & evolution
- Joseph L. Goldstein (1940-), US medical geneticist, Nobel Prize-winner on cholesterol
- Richard M. Goodman (1932-1989), US-Israeli clinical geneticist, pioneered Jewish genetic diseases

- Robert J. Gorlin (1923-2006) US oral pathologist, clinical geneticist, craniofacial syndrome expert
- Carol W. Greider (1961-), US molecular biologist, Lasker Award for telomeres and telomerase
- Frederick Griffith (1879-1941), British medical officer who found transforming principle now called DNA
- Clifford Grobstein (1916-1998), US scientist, bridged classical embryology and developmental biology
- Jean de Grouchy (1926-2003), French pioneer of clinical cytogenetics & karyotype-phenotype correlation
- Hans Gruneberg (1907-1982), British mouse geneticist and blood cell biologist
- Pierre-Henri Gouyon (1953 -), French biologist specializing in genetics and bioethics
- Elliot S. Goldstein American geneticist at Arizona State University
- Ernst Hadorn (1902-1976), Swiss pioneer in developmental genetics, mentor of Walter Gehring
- JBS Haldane (1892-1964), brilliant British human geneticist and co-founder of population genetics
- Ben Hall, US geneticist, DNA:RNA hybridization, yeast production of genetically engineered proteins
- Judy Hall (1939-), dual American and Canadian charismatic clinical geneticist and dysmorphologist
- Dean Hamer (1951-) US geneticist, postulated gay gene and God gene for religious experience
- John Hamerton (1929-2006), Anglo-Canadian cytogeneticist, prenatal diagnostician, bioethicist
- W.D. Hamilton (1936-2000), British evolutionary biologist and eminent evolutionary theorist
- Phil Hanawalt, US geneticist, discovered DNA repair replication
- Anita Harding (1952-1995), UK neurologist, first mitochondrial DNA mutation in disease
- GH Hardy (1877-1947), British mathematician, formulated basic law of population genetics
- Henry Harpending (1944-), US anthropologist and human population geneticist

- Harry Harris (1919-94), British biochemical geneticist par excellence
- Henry Harris (1925-), Australo-British cell biologist, work on cancer and human genetics
- Lee Hartwell (1939-), US yeast geneticist, Nobel Prize, "start" gene and checkpoints in the cell cycle
- Mogens Hauge (1922-1988), Danish medical geneticist and twin researcher
- Donald Hawthorne (1926-2003), US, major contributor to yeast genetics, centromere-linked gene maps
- William Hayes (1918-1994), Australian physician, microbiologist & geneticist, bacterial conjugation
- Robert Haynes (1931-1998), Canadian geneticist and biophysicist, work on DNA repair and mutagenesis
- Frederick Hecht (1930-), US clinical geneticist, cytogeneticist, coined term fragile site
- Michael Heidelberger (1888-1991) US pioneer of modern immunology, won two Lasker Awards
- Martin Heisenberg (1940-), German geneticist,neurobiologist, genetic study of brain of Drosophila
- Charles Roy Henderson, (1911-1989), US animal geneticist, basis for genetic evaluation of livestock
- Al Hershey (1908-1997), US bacterial geneticist, Nobel Prize largely for Hershey-Chase experiment
- Ira Herskowitz (1946–2003), US phage & yeast geneticist, genetic regulatory circuits & mechanisms
- Len Herzenberg (1931-), US human geneticist, immunologist, cell biologist and cell sorter
- Avram Hershko (1937-), Israeli biologist, Nobel Prize for ubiquitin-mediated protein degradation
- Kurt Hirschhorn (1926-), Viennese-born American pediatrician, medical geneticist, cytogeneticist
- Mahlon Hoagland (1921-), US physician and biochemist, co-discovered tRNA with Paul Zamecnik
- Dorothy Hodgkin (1910-1994), British founder of protein crystallography and Nobel Prize winner

- Robert W. Holley (1922-1993), US biochemist, structure of transfer RNA, Nobel Prize
- Leroy Hood (1938-), US molecular biotechnologist, created DNA & protein sequencers & synthesizers
- Norman Horowitz (1915-2005), US geneticist, one gene-one enzyme, chemical evolution, space biology
- H. Robert Horvitz (1947-), US cell biologist, Nobel Prize for programmed cell death
- David E. Housman, US molecular biologist, genetic basis of trinucleotide repeat diseases and cancer
- Martha M. Howe, US phage geneticist, notable contributions to the study of phage Mu
- T.C. Hsu (1917-2003), distinguished Chinese-American cell biologist, geneticist, cytogeneticist
- Thomas J. Hudson (1961-), Canadian genome scientist, maps of human and mouse genomes
- David Hungerford (1927–1993), US co-discoverer of Philadelphia chromosome in CML
- Tim Hunt (1943-), UK biochemist, Nobel Prize for discovery of cyclins in cell cycle control
- Charles Leonard Huskins (1897-1953), English-born Canadian cytogeneticist at McGill and Wisconsin
- Harvey Itano (1925-), American biochemist and pioneer in the study of sickle cell disease
- François Jacob (1920 -), French biologist, won Nobel Prize for bacterial gene control
- Patricia A. Jacobs (1934-), Scottish human geneticist and cytogeneticist
- Albert Jacquard (1925-), French geneticist, essayist, humanist, activist
- Rudolf Jaenisch (1942-), German cell biologist, created transgenic mice, leader in therapeutic cloning
- Richard Jefferson (1956-) US molecular plant biologist in Australia, reporter gene system GUS
- Alec Jeffreys (1950-), British geneticist, developed DNA fingerprinting and DNA profiling techniques

- Niels K. Jerne (1911-1994), Danish, greatest theoretician in modern immunology, Nobel Prize
- Wilhelm Johannsen (1857-1927), Danish botanist who in 1909 coined the word "gene"
- Jonathan D.G. Jones, British plant molecular biologist
- Christian Jung (1956-), German plant geneticist and molecular biologist
- Elvin Kabat (1914–2000) US immunochemist, a founder of modern immunology, antibody-combining sites
- Henrik Kacser (1918-1995), Romanian-born UK biochemist and geneticist, worked on metabolic control
- Axel Kahn (1944-), French scientist and geneticist, known for work on genetically modified plants
- Franz Josef Kallmann (1897-1965), German-US psychiatrist, pioneer in genetics of psychiatric diseases
- Gopinath Kartha (1927-1984), Indian biophysicist, co-discovered triple-helix structure of collagen
- Berwind P. Kaufmann (1897-1975), US botanist, did research in basic plant and animal cytogenetics
- John Kendrew (1917-1997), UK crystallographer, won Nobel Prize for structure of myoglobin
- Cynthia Kenyon (c. 1955-), US molecular biologist, genetics of aging in the worm C. elegans
- Warwick Estevam Kerr (1922-) Brazilian expert in the genetics and sex determination of bees
- Bernard Kettlewell (1907-1979), UK physician, lepidopterist, ecological geneticist, peppered moth
- Seymour Kety (1915-2000), US neuroscientist, essential involvement of genetic factors in schizophrenia
- Gobind Khorana (1922-), Indian-US molecular biologist, synthesized nucleic acids, Nobel Prize
- Motoo Kimura (1924-1994), influential Japanese mathematical biologist in theoretical population genetics
- Mary-Claire King (1946-), US human geneticist and social activist, identified breast cancer genes
- David Klein, (1908-1993), Swiss ophthalmologist and human geneticist

- Harold Klinger (1929-2004), US pioneer on human chromosomes, founded journal *Cytogenetics*
- Aaron Klug (1926-), Lithuania/S Africa/UK, Nobel Prize for developing electron crystallography
- Al Knudson (1922-), US pediatric oncologist, geneticist, formulated two hit hypothesis of cancer
- Georges J. F. Köhler (1946-1995), German, Nobel Prize for hybridomas making monoclonal antibodies
- Arthur Kornberg (1918-), US biochemist, Nobel Prize on DNA synthesis, father of Roger Kornberg
- Roger Kornberg (1947-), US biologist, Nobel Prize on eukaryotic transcription
- Hans Kornberg (1928-), German-UK biologist, studies of carbohydrate transport
- Ed Krebs (1918-), US biochemist, Nobel Prize for phosphorylation as switch activating proteins
- Eric Kremer, US molecular biologist, found trinucleotide repeat in fragile X, research now in gene therapy
- Henry Kunkel (1916–1983), US immunologist, created starch gel electrophoresis to separate proteins
- Bruce Lahn (1969-), Chinese-born geneticist specializing in evolutionary changes of the human brain
- Jean-Baptiste Lamarck (1744-1829), French naturalist, evolutionist, "inheritance of acquired traits"
- Eric Lander (1957-), American molecular geneticist, major contributor to Human Genome Project
- Karl Landsteiner (1868-1943), Austrian-American pathologist, won Nobel Prize for blood group discoveries
- André Langaney, French evolutionary geneticist
- Derald Langham (1913-1991), American agricultural geneticist, the "father of sesame"
- Sam Latt (1938-1988), US pioneer in molecular cytogenetics, fluorescent DNA chromosome probes
- Philip Leder (1934-), US geneticist, method to decode genetic code, transgenic animals to study cancer

- Esther Lederberg (1922-2006), US microbiologist and bacterial genetics pioneer
- Joshua Lederberg (1925-), US molecular biologist, Nobel Prize, headed Rockefeller University
- Jerome Lejeune (1926-1994), French pediatrician, geneticist, discovered trisomy 21 in Down syndrome
- Richard Lenski (1956-), US biologist and phage worker, did long-term E. coli evolution experiment
- Fritz Lenz (1887-1976), German geneticist and eugenicist, ideas influenced Nazi racial hygiene policies
- Widukind Lenz (1919-1995), eminent German medical geneticist who recognized thalidomide syndrome
- Leonard Lerman, US molecular biologist, phage worker, mentor of Nobel Prize-winner Sidney Altman
- Michael Lerner (1910-1977), Russian-US contributor to population, quantitative & evolutionary genetics
- Albert Levan (1905-1998), Swedist geneticist, co-authored report that humans have 46 chromosomes
- Cyrus Levinthal (1922-1990), US molecular geneticist, DNA replication, mRNA, molecular graphics
- Edward B. Lewis (1918-2004), American founder of developmental genetics and Nobel Prize-winner
- Richard Lewontin (1929-), American evolutionary biologist, geneticist and social commentator
- C. C. Li (1912-2003), eminent Chinese American population geneticist and human geneticist
- Wen-Hsiung Li (1942-), Taiwanese-American, molecular evolution, population genetics, genomics
- David Linder (1923-1999), US pathologist and geneticist, used G6PD as X-linked clonal tumor marker
- Susan Lindquist, US molecular biologist studying effects of protein folding and heat-shock proteins
- Jan Lindsten (1935-), eminent Swedish medical geneticist, secretary general of the Nobel Assembly

- Fritz Lipmann (1899-1986), German-American biochemist, Nobel Prize for co-discovery of coenzyme
- C. C. Little (1888–1971), US pioneer mouse geneticist, founded Jackson Laboratory in Bar Harbor
- Richard Losick, US molecular biologist, RNA polymerase, gene transcription, bacterial development
- Herbert Lubs (c.1928-), US internist, medical geneticist, described "marker X" (fragile X chromosome)
- Salvador Luria (1912-1991), Italian-American molecular biologist, Nobel Prize for bacteriophage genetics
- Jay Lush (1896-1982), American animal geneticist who pioneered modern scientific animal breeding
- Michael Lynch, US quantitative geneticist studying evolution, population genetics, and genomics
- Mary F. Lyon (1925-), English mouse geneticist, noted X-inactivation and proposed Lyon hypothesis
- David T. Lykken (1928-2006), American psychologist and behavioral geneticist known for twin studies
- Trofim Lysenko (1898–1976), Soviet scientist, led vicious political campaign against genetics in USSR
- Ellen Magenis (1925-), US medical geneticist and cytogeneticist, Smith-Magenis syndrome
- Phyllis McAlpine (1941-1998), Canadian human geneticist and gene mapper
- Maclyn McCarty (1911–2005), American co-discoverer that DNA is the genetic material
- Barbara McClintock (1902-1992), American cytogeneticist, Nobel Prize for genetic transposition
- W. McGinnis, US molecular geneticist, found homeobox (Hox) genes responsible for basic body plan
- Victor A. McKusick (1921-), US internist and clinical geneticist, organized human genetic knowledge
- Colin MacLeod (1909-1972), Canadian-American co-discoverer that DNA is the genetic material
- Tak Wah Mak (1946-), Chinese-Canadian molecular biologist, co-discovered human T cell receptor genes

- Gustave Malécot (1911-1998), French mathematician who influenced population genetics
- Tom Maniatis (1943-), US molecular biologist, gene cloning, regulation of gene expression
- Clement Markert (1917–1999), eminent US biologist, discovered isozymes
- Joan Marks, American social worker, principal architect of the profession of genetic counselor
- Richard E. Marshall (1933 -), American paediatrician, Greig's syndrome I, Marshall-Smith syndrome
- John Maynard Smith (1920-2004), British evolutionary biologist and population geneticist
- Ernst Mayr (1904-2005), leading German-born American evolutionary biologist
- Peter Medawar (1915-1987), Brazilian-born English scientist, Nobel Prize for immunological tolerance
- Craig C. Mello (1960-), American geneticist, Nobel Prize for discovery of RNA interference
- Gregor Mendel (1822-1884), Bohemian monk who discovered laws of Mendelian inheritance
- Carole Mer h, American geneticist who pioneered DNA typing to differentiate between grape varieties
- Matthew Meselson (1930-), US molecular geneticist, work on DNA replication, recombination, repair
- Peter Michaelis, German plant geneticist, focused on cytoplasmic inheritance
- Ivan Vladimirovich Michurin (1855-1935), Russian plant geneticist, scientific agricultural selection
- Friedrich Mieschler (1844-1895), Swiss biologist, found weak acid in white blood cells now called DNA
- Margareta Mikkelsen (1923–2004), eminent German-born Danish human geneticist and cytogeneticist
- Lois K. Miller (d. 2006, age 54), entomologist and molecular geneticist, studied insect viruses
- O.J. Miller, US physician, human and mammalian genetics and chromosome structure and function
- César Milstein (1927-2002) Argentine-UK, Nobel Prize for hybridomas making monoclonal antibodies

- Aubrey Milunsky (c.1936-), S. African-US physician, medical geneticist, writer, prenatal diagnosis
- Alfred Mirsky (1900—1974), US pioneer in molecular biology, hemoglobin structure, constancy of DNA
- Felix Mitelman, Swedish cancer geneticist and cytogeneticist, catalog of chromosomes in cancer
- Jan Mohr (1921-), eminent Norwegian-Danish pioneer in human gene mapping
- Jacques Monod (1910-1976), French molecular biologist, Nobel Prize-winner
- Lilian Vaughn Morgan (1870-1952), wife of T.H. Morgan and a fine geneticist in her own right
- T.H. Morgan (1866-1945), head of the "fly room," first geneticist to win the Nobel Prize
- Newton E. Morton (1929-), population geneticist and genetic epidemiologist
- Arno G. Motulsky, German-US hematologist who influenced the evolution of medical genetics
- Arthur Mourant (1904-1994), British hematologist, first to examine worldwide blood group distributions
- H.J. Muller (1890-1967) American Drosophila geneticist, Nobel Prize for producing mutations by X-rays
- Hans J. Müller-Eberhard (1927-1998), German-US immunogeneticist, immunoglobulins & complement
- Kary Mullis (1944-), American biochemist, Nobel Prize for the polymerase chain reaction (PCR)
- Walter E. Nance (1933-), US internist and geneticist, research on twins and genetics of deafness
- Daniel Nathans (1928-1999), US microbiologist, Nobel Prize for restriction endonucleases
- James V. Neel (1915-2000), distinguished human geneticist, founded first genetics clinic in the US
- Fred Neidhardt, US microbiologist, pioneer in molecular physiology and proteomics of E. coli
- Oliver Nelson (1920-), US maize geneticist, profound impact on agriculture and basic genetics

- Walter Nelson-Rees, US cytogeneticist, confirmed HeLa cells contamination of other cell lines
- Eugene W. Nester, US microbial geneticist, genetics of Agrobacterium (crown gall formation)
- Carl Neuberg, early pioneer of the study of metabolism.
- Hans Neurath (1909-2002), Austrian-US protein chemist, helped set stage for proteomics
- Marshall W. Nirenberg (1927-), US geneticist, biochemist and Nobel Prize-winner
- Eva Nogales, Spanish biophysicist studying eukaryotic transcription and translation initiation complexes
- Edward Novitski (1918-2006), eminent US Drosophila geneticist, pioneer in chromosome mechanics
- Paul Nurse (1949-), UK biochemist, Nobel Prize for work on CDK, a key regulator of the cell cycle
- Christiane Nüsslein-Volhard (1942-), German developmental biologist and Nobel Prize-winner
- William Nyhan (1926-), US pediatrician and biochemical geneticist, described Lesch-Nyhan syndrome
- Severo Ochoa (1905-1993), Spanish-American biochemist, Nobel Prize for work on the synthesis of RNA
- Susumu Ohno (1928-2000), Japanese-US biologist, evolutionary cytogenetics and molecular evolution
- Tomoko Ohta, Japanese scientist in molecular evolution, the nearly neutral theory of evolution
- Pete Oliver (1898–1991), American geneticist, switched from Drosophila to human genetics
- Jane M. Olson (1952-2004), American genetic epidemiologist and biostatistician
- Maynard Olson, American geneticist, pioneered map of yeast genome and Human Genome Project
- John Opitz (1935-), German-American medical geneticist, expert on dysmorphology and syndromes
- Harry Ostrer, American medical geneticist, studies origins of Jewish peoples
- Ray Owen (1915-), US geneticist, immunologist, found cattle blood groups and chimeric twin calves

- Svante Pääbo (1955-), Swedish molecular anthropologist in Leipzig studying Neanderthal genome
- David Page, US physician and geneticist who mapped, cloned and sequenced the human Y chromosome
- Theophilus Painter (1889-1969), US zoologist, studied fruit fly and human testis chromosomes
- Arthur Pardee (1921-), American scientist who discovered restriction point in the cell cycle
- Klaus Patau (1908–1975), German-American cytogeneticist, described trisomy 13
- Linus Pauling (1901-1994), eminent American chemist, won Nobel Prizes for chemical bonds and peace
- Crodowaldo Pavan (1919-), Brazilian biologist, fly geneticist, and influential scientist in Brazil
- Rose Payne (1909-1999), US transplant geneticist, key to discovery and development of HLA system
- Raymond Pearl (1879-1940), American biologist, biostatistician, rejected eugenics
- Karl Pearson (1857–1936), British statistician, made key contributions to genetic analysis
- LS Penrose (1898-1972), British psychiatrist, human geneticist, pioneered genetics of mental retardation
- Max Perutz (1914-2002), Austrian-British molecular biologist, Nobel Prize for structure of hemoglobin
- Massimo Pigliucci (1964-), Italian-US plant ecological and evolutionary geneticist. Winner of the Dobzhansky Prize.
- Alfred Ploetz (1860-1940), German physician, biologist, eugenicist, introduced racial hygiene to Germany
- Paul Polani (1914-2006), Triese-born UK pediatrician, major catalyst of medical genetics in Britain
- Charles Pomerat (1905–1951), American cell biologist, pioneered the field of tissue culture
- Guido Pontecorvo (1907-1999), Italian-born Scottish geneticist and pioneer molecular biologist
- George R. Price (1922-1975), brilliant but troubled US population geneticist and theoretical biologist

- Peter Propping (1942-), German human geneticist, studies of epilepsy
- Mark Ptashne (c.1940-), US molecular biologist, studies of genetic switch, phage lambda
- Ted Puck (1916–2005), US physicist, work in mammalian & human cell culture, genetics, cytogenetics
- RC Punnett (1875-1967), early English geneticist, discovered linkage with Bateson, stimulated GH Hardy
- Lluis Quintana-Murci (1970-), Spanish human population geneticist, heads part of Genographic Project
- Robert Race (1907-1984), British expert on blood groups, along with wife Ruth Sanger
- Venki Ramakrishnan (c. 1950-), Indian structural biologist, studies of chromatin and ribosome
- Sheldon C. Reed (1910-2003), American pioneer in genetic counseling and behavioral genetics
- G.N. Ramachandran (1922-2001) Indian biophysicist, co-discovered triple-helix structure of collagen
- David Reich, US, human population genetics and genomics, did humans and chimps interbreed?
- Theodore Reich (1938-2003), Canadian-American psychiatrist, a founder of modern psychiatric genetics
- Alexander Rich (1925-), US biologist, biophysicist, discovered Z-DNA and tRNA 3-dimensional structure
- Rollin C. Richmond, US, evolutionary and pharmacogenetic studies of Drosophila, university administrator
- Neil Risch, American human and population geneticist, studied torsion dystonia
- Otto Renner (1883-1960), German plant geneticist, established maternal plastid inheritance
- Marcus Rhoades (1903-1991), great maize (corn) geneticist and cytogeneticist
- David L. Rimoin (1936-), Canadian-US pediatric geneticist, focus on particularly skeletal dysplasias
- Richard Roberts (1943-), British molecular biologist, Nobel Prize for introns and gene-splicing

- Arthur Robinson (1914-2000), American pediatrician, geneticist, pioneer on sex chromosome anomalies
- Herschel L. Roman (1914-1989), American geneticist, innovated in analysis in maize and budding yeast
- Irwin Rose (1926-), American biologist, Nobel Prize for ubiquitin-mediated protein degradation
- Leon Rosenberg (c.1932-), US physician-geneticist, molecular basis of inherited metabolic disease
- Peyton Rous (1879–1970), American tumor virologist and tissue culture expert, Nobel Prize
- Janet Rowley (1925-), American cancer cytogeneticist who found Ph chromosome due to translocation
- Peter T. Rowley (1929–2006), American internist and geneticist, genetics of cancer and leukemia
- Frank Ruddle, US biologist, somatic cell genetics, human gene mapping, paved way for transgenic mice
- Ernst Rüdin (1874-1952), Swiss psychiatrist, geneticist and eugenicist who promoted racial hygiene
- Elizabeth S. Russell (1913-2001), US mammalian geneticist, pioneering work on pigmentation, blood-forming cells, and germ cells
- Liane B. Russell (c. 1923-), Austrian-born US mouse geneticist and radiation biologist
- William L. Russell (1910-2003), UK-US mouse geneticist, pioneered study of mutagenesis in mice
- Leo Sachs (1924-), German-Israeli molecular cancer biologist, colony-stimulating factors, interleukins
- Ruth Sager (1918-1997), US geneticist, pioneer of cytoplasmic genetics, tumor suppressor genes
- Joseph Sambrook (1939-), British viral geneticist
- Avery A. Sandberg, US internist, discovered XYY in 1961, expert on chromosomes in cancer
- Lodewijk A. Sandkuijl (1953-2002), Dutch expert on genetic epidemiology and statistical genetics
- Larry Sandler (1929-1987), US Drosophila geneticist, chromosome mechanics, devoted teacher

- John C. Sanford (1950-), American horticultural geneticist and intelligent design advocate
- Fred Sanger (1918-), UK biochemist, two Nobel Prizes, sequence of insulin, DNA sequencing method
- Ruth Sanger (1918-2001), Australian expert on blood groups, along with husband Robert Race
- Karl Sax (1892-1973), American botanist and cytogeneticist, effects of radiation on chromosomes
- Paul Schedl (1947-), US molecular biologist, genetic regulation of developmental pathways in fruit fly
- Albert Schinzel (1944-), Austrian human geneticist, clinical genetics, karyotype-phenotype correlations
- Werner Schmid (1930-2002), Swiss pioneer in human cytogenetics, described cat-eye syndrome
- Gertrud Schüpbach, Swiss-American biologist, molecular and genetic mechanisms in oogenesis
- Charles Scriver (1930-), Canadian pediatrician, biochemical geneticist, newborn metabolic screening
- Ernie Sears, (1910-1991), Wheat Geneticist who pioneered methods of transferring desirable genes from wild relatives to cultivated wheat in order to increase wheat's resistance to various insects and diseases
- Jay Seegmiller (1920-2006), US human biochemical geneticist, found cause of Lesch-Nyhan syndrome
- Fred Sherman (c. 1933-), US geneticist, one of the "fathers" and mentors of modern yeast genetics
- Larry Shapiro, US pediatric geneticist, lysosomal storage disorders, X chromosome inactivation
- Lucy Shapiro, US molecular geneticist, gene expression during the cell cycle, bacterium Caulobacter
- Phillip Sharp (1944-), US geneticist and molecular biologist, Nobel Prize for co-discovery of gene splicing
- Philip Sheppard (1921–1976), British population geneticist, lepidopterist, human blood group researcher
- G. H. Shull (1874-1954), American geneticist, made key discoveries including heterosis

- Obaid Siddiqui (1932-), Indian neurogeneticist, pioneer on olfactory sense of fruit fly Drosophila
- Norman Simmons (1915-2004), US, forgotten donor of pure DNA to Rosalind Franklin in double helix saga
- Piotr Slonimski (1922-), Polish-Parisian yeast geneticist, pioneer of mitochondrial heredity
- William S. Sly (c. 1931-), US biochemical geneticist, mucopolysaccharidosis type VII (Sly syndrome)
- Cedric A. B. Smith (1917-2002), British statistician, made key contributions to statistical genetics
- David W. Smith (1926-1981), US pediatrician, influential dysmorphologist, named fetal alcohol syndrome
- Hamilton Smith (1931-), American microbiologist, Nobel Prize for restriction endonucleases
- Michael Smith (1932-2000), UK-born Canadian biochemist, Nobel Prize for site-directed mutagenesis
- Oliver Smithies (1925-), UK/US molecular geneticist, inventor, gel electrophoresis, knock-out mice
- George Snell (1903-1996), US mouse geneticist, pioneer transplant immunologist, won Nobel Prize
- Lawrence H. Snyder (1901-1986), American pioneer in medical genetics, studied blood groups
- Robert R. Sokal (1925-), Austrian-born US biological anthropologist and biostatistician.
- Tracy M. Sonneborn (1905–1981), protozoan biologist and geneticist
- Ed Southern (1938-), UK molecular biologist, invented Southern blot and DNA microarray technologies
- Hans Spemann (1869-1941) German embryologist, Nobel Prize for discovery of embryonic induction
- David Stadler, American geneticist, mechanisms of mutation and recombination in Neurospora
- L.J. Stadler (1896-1954), eminent American maize geneticist, father of David Stadler
- Frank Stahl (1929-), American molecular biologist, the Stahl half of the Meselson-Stahl experiment

- David States, US geneticist & bioinformatician, computational study of human genome & proteome
- G. Ledyard Stebbins (1906-2000), American botanist, geneticist and evolutionary biologist
- Michael Stebbins, American geneticist, science writer, or and activist
- Emmy Stein (1879-1954), German botanist and geneticist
- Joan A. Steitz (c.1942-), US molecular biologist, pioneering studies of snRNAs and snRNPs (snurps)
- Gunther Stent (1924-), German-born US molecular geneticist, phage worker, philosopher of science
- Curt Stern (1902-1981), German-born US Drosophila and human geneticist, great teacher
- Nettie Stevens (1861-1912), US geneticist, studied chromosomal basis of sex and discovered XY basis
- Miodrag Stojkoviæ (1964-), Serbian geneticist, working in Europe on mammalian cloning
- George Streisinger (1927-1984), American geneticist, work on bacterial viruses, frameshift mutations
- Alfred Sturtevant (1891–1970), constructed first genetic map of a chromosome
- John Sulston (1942-), British molecular biologist, Nobel Prize for programmed cell death in C. elegans
- James Sumner (1887-1955), American biochemist, Nobel Prize, found enzymes can be crystallized
- Maurice Super (d. 2006, age 69), S. African-born UK pediatric geneticist, studied cystic fibrosis
- Grant Sutherland, Australian molecular cytogeneticist, pioneer on human fragile sites, human genome
- Walter Sutton (1877-1916), US surgeon and scientist, proved chromosomes contained genes
- David Suzuki (1936-), Canadian Drosophila geneticist, science broadcaster and environmental activist
- M.S. Swaminathan (1925-), Indian agricultural scientist, geneticist, leader of Green Revolution in India
- Bryan Sykes, British human geneticist, discovered ways to extract DNA from fossilized bones

- Jack Szostak (1952-), Anglo-US geneticist, work on telomeres, recombination, artificial chromosomes
- Edward Tatum (1909-1975), showed genes control individual steps in metabolism
- Howard Temin (1934-1994), US geneticist, Nobel Prize for discovery of reverse transcriptase
- Alan Templeton (c.1948-), US geneticist & biostatistician, molecular evolution, evolutionary biology
- Donnall Thomas (1920-) US physician, Nobel Prize for bone marrow transplantation for leukemia
- Nikolay Timofeeff-Ressovsky (1900-1981), Russian radiation and evolution geneticist
- Alfred Tissières (1917-2003), Swiss molecular geneticist, pioneered molecular biology in Geneva
- Joe Hin Tjio (1919-2001), Java-born geneticist who first discovered humans have 46 chromosomes
- Susumu Tonegawa (1939-), Japanese molecular biologist, Nobel Prize for genetics of antibody diversity
- Erich von Tschermak (1871-1962) Austrian agronomist and one of the re-discoverers of Mendel's laws
- Lap-Chee Tsui, Chinese geneticist, sequenced first human gene (for cystic fibrosis) with Collins
- Raymond Turpin (1895-1988), French pediatrician, geneticist, Lejeune's co-discoverer of trisomy 21
- Axel Ullrich (1943-), German molecular biologist, signal transduction, discovered oncogene, Herceptin
- Irene Ayako Uchida (1917-), Canadian geneticist and cytogeneticist. One of the first in Canada. Down syndrome
- Harold Varmus (1939-), American Nobel Prize-winner for oncogenes, head of NIH
- Nikolai Vavilov (1887-1943), eminent Russian botanist and geneticist, anti-Lysenko, died in prison
- Craig Venter (1946-), American molecular biologist and entrepreneur, raced to sequence the genome

- Jerome Vinograd (1913-1976), US, leader in biochemistry and molecular biology of nucleic acids
- Friedrich Vogel, German, leader in human genetics, coined term "pharmacogenetics"
- Bert Vogelstein (1949-), US pediatrician and cancer geneticist, series of mutations in colorectal cancer
- Erik Adolf von Willebrand (1870-1949), Finnish internist who found commonest bleeding disorder
- Petrus Johannes Waardenburg (1886-1979), Dutch ophthalmologist, geneticist, Waardenburg syndrome
- C.H. Waddington (1905-1975), British developmental biologist, paleontologist, geneticist, embryologist
- Alfred Russel Wallace (1823-1913), Welsh, proposed natural selection theory independent of Darwin
- Douglas Wallace, US mitochondrial geneticist, pioneered human mtDNA as a molecular marker
- Peter Walter, German-US molecular biologist studying protein folding and protein targeting
- Richard H. Ward (1943-2003), English-born New Zealand human and anthropological geneticist
- James D. Watson (1928-), US molecular geneticist, Nobel Prize for discovery of the double helix
- David Weatherall, distinguished UK physician, geneticist, pioneer in hemoglobin and molecular medicine
- Robert Weinberg, US, discovered first human oncogene and first tumor suppressor gene
- Wilhelm Weinberg (1862-1937),German physician, formulated basic law of population genetics
- Spencer Wells (1969-), US genetic anthropologist, head of Genographic Project to map past migrations
- Susan R. Wessler (1953-), US plant molecular geneticist, transposable elements re genetic diversity
- Raymond White, US cancer geneticist, cloned APC colon cancer gene & neurofibromatosis gene

- Glayde Whitney (1939–2002) US behavioral geneticist, accused of supporting scientific racism
- Reed Wickner (c. 1940-) US molecular geneticist, yeast phenotypes due to prion forms of native proteins
- Alexander S. Wiener (1907-1976), U.S. immunologist, discovered Rh blood groups with Landsteiner
- Eric F. Wieschaus (1947-), American developmental biologist and Nobel Prize-winner
- Maurice Wilkins (1916-2004), New Zealand-born British Nobel Prize-winner with Watson and Crick
- Huntington Willard (c.1953-), US human geneticist, X chromosome inactivation, gene silencing
- Robley Williams (1908-1995), US virologist, recreated tobacco mosaic virus from its RNA + protein coat
- Ian Wilmut (1944-) UK reproductive biologist who first cloned a mammal (lamb named Dolly)
- Allan Wilson (1934-1991) New Zealand-US innovator in molecular study of human evolution
- David Sloan Wilson (1949-), US evolutionary biologist and geneticist
- Edmund Beecher Wilson (1856-1939), US zoologist, geneticist, discovered XY & XX sex chromosomes
- Øjvind Winge (1886-1964), Danish biologist and pioneer in yeast genetics
- Chester B. Whitley (1950-), US geneticist, pioneered treatment of lysosomal diseases
- Carl Woese (1928-), US biologist, defined Archaea as new domain of life, rRNA phylogenetic tool
- Ulrich Wolf (1933-), German cytogeneticist, found chromosome 4p deletion in Wolf-Hirschhorn syndrome
- Melaku Worede (1936-), Ethiopian conservationist and geneticist
- Sewall Wright (1889–1988), eminent US geneticist who, with Fisher, united genetics & evolution
- Charles Yanofsky (1925-), American molecular geneticist, colinearity of gene and its protein product
- Floyd Zaiger (1926-), fruit geneticist and entrepreneur

- Hans Zellweger (1909-1990) Swiss-US pediatrician and clinical geneticist, described Zellweger syndrome
- Norton Zinder (1928-) American biologist and phage worker who discovered genetic transduction
- Rolf M. Zinkernagel (1944-), Swiss scientist, won Nobel Prize for immune recognition of antigen

REFERENCES

A.J.F. Griffiths, J.H. Miller, D.T. Suzuki, R.C. Lewontin, and W.M. Gelbart (2000). An Introduction to Genetic Analysis. W.H. Freeman and Company. Chapter 8 (The Structure and Replication of DNA): Mechanism of DNA Replication.

A.J.F. Griffiths, J.H. Miller, D.T. Suzuki, R.C. Lewontin, and W.M. Gelbart (2000). An Introduction to Genetic Analysis. W.H. Freeman and Company. Chapter 3 (Chromosomal Basis of Heredity): Mendelian genetics in eukaryotic life cycles.

A.J.F. Griffiths, J.H. Miller, D.T. Suzuki, R.C. Lewontin, and W.M. Gelbart (2000). An Introduction to Genetic Analysis. W.H. Freeman and Company. Chapter 2 (Patterns of Inheritance): Sex chromosomes and sex-linked inheritance.

A.J.F. Griffiths, J.H. Miller, D.T. Suzuki, R.C. Lewontin, and W.M. Gelbart (2000). An Introduction to Genetic Analysis. W.H. Freeman and Company. Chapter 7 (Gene Transfer in Bacteria and Their Viruses): Bacterial conjugation.

A.J.F. Griffiths, J.H. Miller, D.T. Suzuki, R.C. Lewontin, and W.M. Gelbart (2000). An Introduction to Genetic Analysis. W.H. Freeman and Company. Chapter 7 (Gene Transfer in Bacteria and Their Viruses): Bacterial transformation.

A.J.F. Griffiths, J.H. Miller, D.T. Suzuki, R.C. Lewontin, and W.M. Gelbart (2000). An Introduction to Genetic Analysis. W.H. Freeman and Company. Chapter 5 (Basic Eukaryotic Chromosome Mapping): Nature of crossing-over.

A.J.F. Griffiths, J.H. Miller, D.T. Suzuki, R.C. Lewontin, and W.M. Gelbart (2000). An Introduction to Genetic Analysis. W.H. Freeman and Company. Chapter 5 (Basic Eukaryotic Chromosome Mapping): Linkage maps.

Alberts B, Johnson A, Lewis J, Raff M, Roberts K, and Walter P (2002). Molecular Biology of the Cell, 4th ion. DNA and chromosomes: Chromosomal DNA and Its Packaging in the Chromatin Fiber.

Alberts B, Johnson A, Lewis J, Raff M, Roberts K, and Walter P (2002). Molecular Biology of the Cell, 4th ion. Proteins: The Shape and Structure of Proteins.

Alberts B, Johnson A, Lewis J, Raff M, Roberts K, and Walter P (2002). Molecular Biology of the Cell, 4th ion. Proteins: Protein Function.

Alberts B, Johnson A, Lewis J, Raff M, Roberts K, and Walter P (2002). Molecular Biology of the Cell, 4th ion. Control of Gene Expression - The Tryptophan Repressor Is a Simple Switch That Turns Genes On and Off in Bacteria.

Avery OT, MacLeod CM, and McCarty M (1944). "Studies on the Chemical Nature of the

Substance Inducing Transformation of Pneumococcal Types: Induction of Transformation by a Desoxyribonucleic Acid Fraction Isolated from Pneumococcus Type III". Journal of Experimental Medicine 79 (1): 137-58. 35th anniversary reprint available.

Bateson, W (1907). "The Progress of Genetic Research". Wilks, W (or) Report of the Third 1906 International Conference on Genetics: Hybridization (the cross-breeding of genera or species), the cross-breeding of varieties, and general plant breeding, London: Royal Horticultural Society.

Brivanlou AH, Darnell JE Jr (2002). "Signal transduction and the control of gene expression". Science 295 (5556): 813-8.

D.L. Imes, L.A. Geary, R.A. Grahn, L.A. Lyons (2006). "Albinism in the domestic cat (Felis catus) is associated with a tyrosinase (TYR) mutation" (Short Communication). Animal Genetics 37 (2): 175. Retrieved on 2006-05-29.

Ernest W. Crow and James F. Crow (2002). "100 Years Ago: Walter Sutton and the Chromosome Theory of Heredity". Genetics 160.

Fiers W et al., Complete nucleotide-sequence of bacteriophage MS2-RNA - primary and secondary structure of replicase gene, Nature, 260, 500-507, 1976

Griffiths AJF, Miller JH, Suzuki DT, Lewontin RC, and Gelbart WM (2000). An Introduction to Genetic Analysis. W.H. Freeman and Company. Chapter 2 (Patterns of Inheritance): Mendel's experiments

H. Lodish, A. Berk, L.S. Zipursky, P. Matsudaira, D. Baltimore, and J. Darnell (2000). Molecular Cell Biology, 4th ion. Chapter 7 (Recombinant DNA and Genomics): 7.1. DNA Cloning with Plasmid Vectors

Henig, Robin Marantz (2000). The Monk in the Garden : The Lost and Found Genius of Gregor Mendel, the Father of Genetics. Houghton Mifflin. ISBN 0-395-97765-7. "The article, written by an obscure Moravian monk named Gregor Mendel..."

J.M. Berg, J.L. Tymoczko, L. Stryer, N.D. Clarke (2002). Biochemistry, 5th ion, New York: W. H. Freeman and Company. I. 5. DNA, RNA, and the Flow of Genetic Information: Amino Acids Are Encoded by Groups of Three Bases Starting from a Fixed Point

Jaenisch R, Bird A. "Epigenetic regulation of gene expression: how the genome integrates intrinsic and environmental signals". Nature Genetics 33 (3s): 245-254.

Kenneth R. Bridges, M.D. (2002-04-11). How Does Sickle Cell Cause Disease?. Retrieved on 2007-07-23.

Kunkel TA (2004). "DNA Replication Fidelity". Journal of Biological Chemistry 279 (17): 16895-16898.

Mayeux R (2005). "Mapping the new frontier: complex genetic disorders". The Journal of Clinical Investigation 115 (6): 1404-7.

Mendel, GJ (1866). "Versuche über Pflanzen-Hybriden". Verhandlungen des naturforschenden Vereins Brünn 4: 3-47. (in English in 1901, J. R. Hortic. Soc. 26: 1–32) translation available online

Min Jou W, Haegeman G, Ysebaert M, Fiers W., Nucleotide sequence of the gene coding for the bacteriophage MS2 coat protein, Nature. 1972 May 12;237(5350):82-8

Pearson H (2006). "Genetics: what is a gene?". Nature 441 (7092): 398-401.

Saiki RK, Scharf S, Faloona F, Mullis KB, Horn GT, Erlich HA, Arnheim N (1985). "Enzymatic Amplification of â-Globin Genomic Sequences and Restriction Site Analysis for Diagnosis of Sickle Cell Anemia". Science 230 (4732): 1350-1354.

Sanger F, Air GM, Barrell BG, Brown NL, Coulson AR, Fiddes CA, Hutchison CA, Slocombe PM, Smith M., Nucleotide sequence of bacteriophage phi X174 DNA, Nature. 1977 Feb 24;265(5596):687-95

Sanger F, Nicklen S, and Coulson AR (1977). "DNA sequencing with chain-terminating inhibitors". Nature 74 (12): 5463-5467.

T. Strachan and A.P. Read (1999). Human Molecular Genetics 2, second ion, John Wiley & Sons Inc.. Chapter 18: Cancer Genetics

V.L. Chandler (2007). "Paramutation: From Maize to Mice". Cell 128: 641-645.

Vries, H. de (1889) Intracellular Pangenesis ("pan-gene" definition on page 7 and 40 of this 1910 translation in English)

Watson JD, Crick FH, Molecular structure of nucleic acids; a structure for deoxyribose nucleic acid, Nature. 1953 Apr 25;171(4356):737-8

Weiling F (1991). "Historical study: Johann Gregor Mendel 1822-1884". American Journal of Medical Genetics 40 (1): 1-25; discussion 26. PMID 1887835.

2

Elements of Genetic Code, Gene Expression & Evolution

DNA

Deoxyribonucleic acid (DNA) is a nucleic acid that contains the genetic instructions used in the development and functioning of all known living organisms. The main role of DNA molecules is the long-term storage of information and DNA is often compared to a set of blueprints, since it contains the instructions needed to construct other components of cells, such as proteins and RNA molecules. The DNA segments that carry this genetic information are called genes, but other DNA sequences have structural purposes, or are involved in regulating the use of this genetic information.

Chemically, DNA is a long polymer of simple units called nucleotides, with a backbone made of sugars and phosphate groups joined by ester bonds. Attached to each sugar is one of four types of molecules called bases. It is the sequence of these four bases along the backbone that encodes information. This information is read using the genetic code, which specifies the sequence of the amino acids within proteins. The code is read by copying stretches of DNA into the related nucleic acid RNA, in a process called transcription. Most of these RNA molecules are used to synthesize proteins, but others are used directly in structures such as ribosomes and spliceosomes.

Within cells, DNA is organized into structures called chromosomes. These chromosomes are duplicated before cells divide, in a process called DNA replication. Eukaryotic organisms such as animals, plants, and fungi store their DNA inside the cell nucleus, while in prokaryotes such as bacteria it is found in the cell's cytoplasm. Within the chromosomes, chromatin proteins such as histones compact and organize

DNA, which helps control its interactions with other proteins and thereby control which genes are transcribed.

Physical and Chemical Properties

DNA is a long polymer made from repeating units called nucleotides. The DNA chain is 22 to 26 Ångströms wide (2.2 to 2.6 nanometres), and one nucleotide unit is 3.3 Ångstroms (0.33 nanometres) long. Although each individual repeating unit is very small, DNA polymers can be enormous molecules containing millions of nucleotides. For instance, the largest human chromosome, chromosome number 1, is 220 million base pairs long.

In living organisms, DNA does not usually exist as a single molecule, but instead as a tightly-associated pair of molecules. These two long strands entwine like vines, in the shape of a double helix. The nucleotide repeats contain both the segment of the backbone of the molecule, which holds the chain together, and a base, which interacts with the other DNA strand in the helix. In general, a base linked to a sugar is called a nucleoside and a base linked to a sugar and one or more phosphate groups is called a nucleotide. If multiple nucleotides are linked together, as in DNA, this polymer is referred to as a polynucleotide.

The backbone of the DNA strand is made from alternating phosphate and sugar residues. The sugar in DNA is 2-deoxyribose, which is a pentose (five carbon) sugar. The sugars are joined together by phosphate groups that form phosphodiester bonds between the third and fifth carbon atoms of adjacent sugar rings. These asymmetric bonds mean a strand of DNA has a direction. In a double helix the direction of the nucleotides in one strand is opposite to their direction in the other strand. This arrangement of DNA strands is called antiparallel. The asymmetric ends of DNA strands are referred to as the 52 (*five prime*) and 32 (*three prime*) ends. One of the major differences between DNA and RNA is the sugar, with 2-deoxyribose being replaced by the alternative pentose sugar ribose in RNA.

The DNA double helix is stabilized by hydrogen bonds between the bases attached to the two strands. The four bases found in DNA are adenine (abbreviated A), cytosine (C), guanine (G) and thymine (T). These four bases are shown below and are attached to the sugar/phosphate to form the complete nucleotide, as shown for adenosine monophosphate.

These bases are classified into two types; adenine and guanine are fused five- and six-membered heterocyclic compounds called purines, while cytosine and thymine are six-membered rings called pyrimidines. A fifth pyrimidine base, called uracil (U), usually takes the place of thymine in RNA and differs from thymine by lacking a methyl group on its ring. Uracil is not usually found in DNA, occurring only as a breakdown product of cytosine, but a very rare exception to this rule is a bacterial virus called PBS1 that contains uracil in its DNA. In contrast, following synthesis of certain RNA molecules, a significant number of the uracils are converted to thymines by the enzymatic addition of the missing methyl group. This occurs mostly on structural and enzymatic RNAs like transfer RNAs and ribosomal RNA.

Major and Minor Grooves

The double helix is a right-handed spiral. As the DNA strands wind around each other, they leave gaps between each set of phosphate backbones, revealing the sides of the bases inside (see animation). There are two of these grooves twisting around the surface of the double helix: one groove, the major groove, is 22 Å wide and the other, the minor groove, is 12 Å wide. The narrowness of the minor groove means that the edges of the bases are more accessible in the major groove. As a result, proteins like transcription factors that can bind to specific sequences in double-stranded DNA usually make contacts to the sides of the bases exposed in the major groove.

Base Pairing

Each type of base on one strand forms a bond with just one type of base on the other strand. This is called complementary base pairing. Here, purines form hydrogen bonds to pyrimidines, with A bonding only to T, and C bonding only to G. This arrangement of two nucleotides binding together across the double helix is called a base pair. In a double helix, the two strands are also held together via forces generated by the hydrophobic effect and pi stacking, which are not influenced by the sequence of the DNA. As hydrogen bonds are not covalent, they can be broken and rejoined relatively easily. The two strands of DNA in a double helix can therefore be pulled apart like a zipper, either by a mechanical force or high temperature. As a result of this complementarity, all the information in the double-stranded sequence of a DNA helix is duplicated on each strand, which is vital in DNA replication. Indeed, this reversible and specific interaction between complementary base pairs is critical for all the functions of DNA in living organisms.

The two types of base pairs form different numbers of hydrogen bonds, AT forming two hydrogen bonds, and GC forming three hydrogen bonds (see figures, left). The GC base pair is therefore stronger than the AT base pair. As a result, it is both the percentage of GC base pairs and the overall length of a DNA double helix that determine the strength of the association between the two strands of DNA. Long DNA helices with a high GC content have stronger-interacting strands, while short helices with high AT content have weaker-interacting strands. Parts of the DNA double helix that need to separate easily, such as the TATAAT Pribnow box in bacterial promoters, tend to have sequences with a high AT content, making the strands easier to pull apart. In the laboratory, the strength of this interaction can be measured by finding the temperature required to break the hydrogen bonds, their melting temperature (also called T_m value). When all the base pairs in a DNA double helix melt, the strands separate and exist in solution as two entirely independent molecules. These single-stranded DNA molecules have no single common shape, but some conformations are more stable than others.

Sense and Antisense

A DNA sequence is called "sense" if its sequence is the same as that of a messenger RNA copy that is translated into protein. The sequence on the opposite strand is complementary to the sense sequence and is therefore called the "antisense" sequence. Since RNA polymerases work by making a complementary copy of their templates, it is this antisense strand that is the template for producing the sense messenger RNA. Both sense and antisense sequences can exist on different parts of the same strand of DNA (i.e. both strands contain both sense and antisense sequences). In both prokaryotes and eukaryotes, antisense RNA sequences are produced, but the functions of these RNAs are not entirely clear. One proposal is that antisense RNAs are involved in regulating gene expression through RNA-RNA base pairing.

A few DNA sequences in prokaryotes and eukaryotes, and more in plasmids and viruses, blur the distinction made above between sense and antisense strands by having overlapping genes. In these cases, some DNA sequences do double duty, encoding one protein when read 52 to 32 along one strand, and a second protein when read in the opposite direction (still 52 to 32) along the other strand. In bacteria, this overlap may be involved in the regulation of gene transcription, while in viruses, overlapping genes increase the amount of information that can be encoded within the small viral genome. Another way of reducing genome size is seen in some viruses that contain linear or circular single-stranded DNA as their genetic material.

Supercoiling

DNA can be twisted like a rope in a process called DNA supercoiling. With DNA in its "relaxed" state, a strand usually circles the axis of the double helix once every 10.4 base pairs, but if the DNA is twisted the strands become more tightly or more loosely wound. If the DNA is twisted in the direction of the helix, this is positive supercoiling, and the bases are held more tightly together. If they are twisted in the opposite direction, this is negative supercoiling, and the bases come apart more easily. In nature, most DNA has slight negative supercoiling that is introduced by enzymes called topoisomerases. These enzymes are also needed to relieve the twisting stresses introduced into DNA strands during processes such as transcription and DNA replication.

Alternative Double-Helical Structures

DNA exists in several possible conformations. The conformations so far identified are: A-DNA, B-DNA, C-DNA, D-DNA, E-DNA, H-DNA, L-DNA, P-DNA, and Z-DNA. However, only A-DNA, B-DNA, and Z-DNA have been observed in naturally occurring biological systems. Which conformation DNA adopts depends on the sequence of the DNA, the amount and direction of supercoiling, chemical modifications of the bases and also solution conditions, such as the concentration of metal ions and polyamines. Of these three conformations, the "B" form described above is most common under the conditions found in cells. The two alternative double-helical forms of DNA differ in their geometry and dimensions.

The A form is a wider right-handed spiral, with a shallow and wide minor groove and a narrower and deeper major groove. The A form occurs under non-physiological conditions in dehydrated samples of DNA, while in the cell it may be produced in hybrid pairings of DNA and RNA strands, as well as in enzyme-DNA complexes. Segments of DNA where the bases have been chemically-modified by methylation may undergo a larger change in conformation and adopt the Z form. Here, the strands turn about the helical axis in a left-handed spiral, the opposite of the more common B form. These unusual structures can be recognised by specific Z-DNA binding proteins and may be involved in the regulation of transcription.

Quadruplex Structures

At the ends of the linear chromosomes are specialized regions of DNA called telomeres. The main function of these regions is to allow the cell to replicate chromosome ends using the enzyme telomerase, as the enzymes that normally replicate DNA cannot copy the extreme 32 ends of chromosomes. As a result, if a chromosome lacked telomeres it would become shorter each time it was replicated. These specialized chromosome caps also help protect the DNA ends from exonucleases and stop the DNA repair systems in the cell from treating them as damage to be corrected. In human cells, telomeres are usually lengths of single-stranded DNA containing several thousand repeats of a simple TTAGGG sequence.

These guanine-rich sequences may stabilize chromosome ends by forming very unusual structures of stacked sets of four-base units, rather than the usual base pairs found in other DNA molecules. Here, four guanine bases form a flat plate and these flat four-base units then stack on top of each other, to form a stable *G-quadruplex* structure. These structures are stabilized by hydrogen bonding between the edges of the bases and chelation of a metal ion in the centre of each four-base unit. The structure shown to the left is a top view of the quadruplex formed by a DNA sequence found in human telomere repeats. The single DNA strand forms a loop, with the sets of four bases stacking in a central quadruplex three plates deep.

In the space at the centre of the stacked bases are three chelated potassium ions. Other structures can also be formed, with the central set of four bases coming from either a single strand folded around the bases, or several different parallel strands, each contributing one base to the central structure.

In addition to these stacked structures, telomeres also form large loop structures called telomere loops, or T-loops. Here, the single-stranded DNA curls around in a long circle stabilized by telomere-binding proteins. At the very end of the T-loop, the single-stranded telomere DNA is held onto a region of double-stranded DNA by the telomere strand disrupting the double-helical DNA and base pairing to one of the two strands. This triple-stranded structure is called a displacement loop or D-loop.

Chemical Modifications

Base Modifications

The expression of genes is influenced by the chromatin structure of a chromosome and regions of heterochromatin (low or no gene expression) correlate with the methylation of cytosine. For example, cytosine methylation, to produce 5-methylcytosine, is important for X-chromosome inactivation. The average level of methylation varies between organisms, with *Caenorhabditis elegans* lacking cytosine methylation, while vertebrates show higher levels, with up to 1% of their DNA containing 5-methylcytosine. Despite the biological role of 5-methylcytosine it is susceptible to spontaneous deamination to leave the thymine base, and methylated cytosines are therefore mutation hotspots. Other base modifications include adenine methylation in bacteria and the glycosylation of uracil to produce the "J-base" in kinetoplastids.

DNA Damage

DNA can be damaged by many different sorts of mutagens. These include oxidizing agents, alkylating agents and also high-energy electromagnetic radiation such as ultraviolet light and x-rays. The type of DNA damage produced depends on the type of mutagen. For example, UV light mostly damages DNA by producing thymine dimers, which are cross-links between adjacent pyrimidine bases in a DNA strand. On the other hand, oxidants such as free radicals or hydrogen peroxide produce multiple forms of damage, including base modifications, particularly of guanosine, as well as double-strand breaks. It has been estimated that in each human cell, about 500 bases suffer oxidative damage per day. Of these oxidative lesions, the most dangerous are double-strand breaks, as these lesions are difficult to repair and can produce point mutations, insertions and deletions from the DNA sequence, as well as chromosomal translocations.

Many mutagens intercalate into the space between two adjacent base pairs.

Intercalators are mostly aromatic and planar molecules, and include ethidium, daunomycin, doxorubicin and thalidomide. In order for an intercalator to fit between base pairs, the bases must separate, distorting the DNA strands by unwinding of the double helix. These structural changes inhibit both transcription and DNA replication, causing toxicity and mutations. As a result, DNA intercalators are often carcinogens, with benzopyrene diol epoxide, acridines, aflatoxin and ethidium bromide being well-known examples. Nevertheless, due to their properties of inhibiting DNA transcription and replication, they are also used in chemotherapy to inhibit rapidly-growing cancer cells.

Overview of Biological Functions

DNA usually occurs as linear chromosomes in eukaryotes, and circular chromosomes in prokaryotes. The set of chromosomes in a cell makes up its genome; the human genome has approximately 3 billion base pairs of DNA arranged into 46 chromosomes. The information carried by DNA is held in the sequence of pieces of DNA called genes. Transmission of genetic information in genes is achieved via complementary base pairing. For example, in transcription, when a cell uses the information in a gene, the DNA sequence is copied into a complementary RNA sequence through the attraction between the DNA and the correct RNA nucleotides. Usually, this RNA copy is then used to make a matching protein sequence in a process called translation which depends on the same interaction between RNA nucleotides. Alternatively, a cell may simply copy its genetic information in a process called DNA replication. The details of these functions are covered in other articles; here we focus on the interactions between DNA and other molecules that mediate the function of the genome.

Genome Structure

Genomic DNA is located in the cell nucleus of eukaryotes, as well as small amounts in mitochondria and chloroplasts. In prokaryotes, the DNA is held within an irregularly shaped body in the cytoplasm called the nucleoid. The genetic information in a genome is held within genes. A gene is a unit of heredity and is a region of DNA that influences a particular characteristic in an organism. Genes contain an open reading frame that can be transcribed, as well as regulatory sequences such as promoters and enhancers, which control the expression of the open reading frame.

In many species, only a small fraction of the total sequence of the genome encodes protein. For example, only about 1.5% of the human genome consists of protein-coding exons, with over 50% of human DNA consisting of non-coding repetitive sequences. The reasons for the presence of so much non-coding DNA in eukaryotic genomes and the extraordinary differences in genome size, or *C-value*, among species represent a long-standing puzzle known as the "C-value enigma." However,

DNA sequences that do not code protein may still encode functional non-coding RNA molecules, which are involved in the regulation of gene expression.

Some non-coding DNA sequences play structural roles in chromosomes. Telomeres and centromeres typically contain few genes, but are important for the function and stability of chromosomes. An abundant form of non-coding DNA in humans are pseudogenes, which are copies of genes that have been disabled by mutation. These sequences are usually just molecular fossils, although they can occasionally serve as raw genetic material for the creation of new genes through the process of gene duplication and divergence.

Transcription and Translation

A gene is a sequence of DNA that contains genetic information and can influence the phenotype of an organism. Within a gene, the sequence of bases along a DNA strand defines a messenger RNA sequence, which then defines one or more protein sequences. The relationship between the nucleotide sequences of genes and the amino-acid sequences of proteins is determined by the rules of translation, known collectively as the genetic code. The genetic code consists of three-letter 'words' called *codons* formed from a sequence of three nucleotides (e.g. ACT, CAG, TTT).

In transcription, the codons of a gene are copied into messenger RNA by RNA polymerase. This RNA copy is then decoded by a ribosome that reads the RNA sequence by base-pairing the messenger RNA to transfer RNA, which carries amino acids. Since there are 4 bases in 3-letter combinations, there are 64 possible codons (4^3 combinations). These encode the twenty standard amino acids, giving most amino acids more than one possible codon. There are also three 'stop' or 'nonsense' codons signifying the end of the coding region; these are the TAA, TGA and TAG codons.

Replication

Cell division is essential for an organism to grow, but when a cell divides it must replicate the DNA in its genome so that the two daughter cells have the same genetic information as their parent. The double-stranded structure of DNA provides a simple mechanism for DNA replication. Here, the two strands are separated and then each strand's complementary DNA sequence is recreated by an enzyme called DNA polymerase. This enzyme makes the complementary strand by finding the correct base through complementary base pairing, and bonding it onto the original strand. As DNA polymerases can only extend a DNA strand in a 52 to 32 direction, different mechanisms are used to copy the antiparallel strands of the double helix. In this way, the base on the old strand dictates which base appears on the new strand, and the cell ends up with a perfect copy of its DNA.

Interactions with Proteins

All the functions of DNA depend on interactions with proteins. These protein interactions can be non-specific, or the protein can bind specifically to a single DNA sequence. Enzymes can also bind to DNA and of these, the polymerases that copy the DNA base sequence in transcription and DNA replication are particularly important.

DNA-Binding Proteins

Structural proteins that bind DNA are well-understood examples of non-specific DNA-protein interactions. Within chromosomes, DNA is held in complexes with structural proteins. These proteins organize the DNA into a compact structure called chromatin. In eukaryotes this structure involves DNA binding to a complex of small basic proteins called histones, while in prokaryotes multiple types of proteins are involved. The histones form a disk-shaped complex called a nucleosome, which contains two complete turns of double-stranded DNA wrapped around its surface. These non-specific interactions are formed through basic residues in the histones making ionic bonds to the acidic sugar-phosphate backbone of the DNA, and are therefore largely independent of the base sequence. Chemical modifications of these basic amino acid residues include methylation, phosphorylation and acetylation. These chemical changes alter the strength of the interaction between the DNA and the histones, making the DNA more or less accessible to transcription factors and changing the rate of transcription. Other non-specific DNA-binding proteins found in chromatin include the high-mobility group proteins, which bind preferentially to bent or distorted DNA. These proteins are important in bending arrays of nucleosomes and arranging them into more complex chromatin structures.

A distinct group of DNA-binding proteins are the single-stranded-DNA-binding proteins that specifically bind single-stranded DNA. In humans, replication protein A is the best-characterised member of this family and is essential for most processes where the double helix is separated, including DNA replication, recombination and DNA repair. These binding proteins seem to stabilize single-stranded DNA and protect it from forming stem loops or being degraded by nucleases.

In contrast, other proteins have evolved to specifically bind particular DNA sequences. The most intensively studied of these are the various classes of transcription factors, which are proteins that regulate transcription. Each one of these proteins bind to one particular set of DNA sequences and thereby activates or inhibits the transcription of genes with these sequences close to their promoters. The transcription factors do this in two ways. Firstly, they can bind the RNA polymerase responsible for transcription, either directly or through other mediator proteins; this locates the polymerase at the promoter and allows it to begin transcription. Alternatively, transcription factors can bind enzymes that modify the histones at the promoter; this will change the accessibility of the DNA template to the polymerase.

As these DNA targets can occur throughout an organism's genome, changes in the activity of one type of transcription factor can affect thousands of genes. Consequently, these proteins are often the targets of the signal transduction processes that mediate responses to environmental changes or cellular differentiation and development. The specificity of these transcription factors' interactions with DNA come from the proteins making multiple contacts to the edges of the DNA bases, allowing them to "read" the DNA sequence. Most of these base-interactions are made in the major groove, where the bases are most accessible.

DNA-Modifying Enzymes

Nucleases and Ligases

Nucleases are enzymes that cut DNA strands by catalyzing the hydrolysis of the phosphodiester bonds. Nucleases that hydrolyse nucleotides from the ends of DNA strands are called exonucleases, while endonucleases cut within strands. The most frequently-used nucleases in molecular biology are the restriction endonucleases, which cut DNA at specific sequences. For instance, the EcoRV enzyme shown to the left recognizes the 6-base sequence 52 -GAT|ATC-32 and makes a cut at the vertical line. In nature, these enzymes protect bacteria against phage infection by digesting the phage DNA when it enters the bacterial cell, acting as part of the restriction modification system. In technology, these sequence-specific nucleases are used in molecular cloning and DNA fingerprinting.

Enzymes called DNA ligases can rejoin cut or broken DNA strands, using the energy from either adenosine triphosphate or nicotinamide adenine dinucleotide. Ligases are particularly important in lagging strand DNA replication, as they join together the short segments of DNA produced at the replication fork into a complete copy of the DNA template. They are also used in DNA repair and genetic recombination.

Topoisomerases and Helicases

Topoisomerases are enzymes with both nuclease and ligase activity. These proteins change the amount of supercoiling in DNA. Some of these enzyme work by cutting the DNA helix and allowing one section to rotate, thereby reducing its level of supercoiling; the enzyme then seals the DNA break. Other types of these enzymes are capable of cutting one DNA helix and then passing a second strand of DNA through this break, before rejoining the helix. Topoisomerases are required for many processes involving DNA, such as DNA replication and transcription.

Helicases are proteins that are a type of molecular motor. They use the chemical energy in nucleoside triphosphates, predominantly ATP, to break hydrogen bonds between bases and unwind the DNA double helix into single strands. These enzymes are essential for most processes where enzymes need to access the DNA bases.

Polymerases

Polymerases are enzymes that synthesise polynucleotide chains from nucleoside triphosphates. They function by adding nucleotides onto the 32 hydroxyl group of the previous nucleotide in the DNA strand. As a consequence, all polymerases work in a 52 to 32 direction. In the active site of these enzymes, the nucleoside triphosphate substrate base-pairs to a single-stranded polynucleotide template: this allows polymerases to accurately synthesise the complementary strand of this template. Polymerases are classified according to the type of template that they use.

In DNA replication, a DNA-dependent DNA polymerase makes a DNA copy of a DNA sequence. Accuracy is vital in this process, so many of these polymerases have a proofreading activity. Here, the polymerase recognizes the occasional mistakes in the synthesis reaction by the lack of base pairing between the mismatched nucleotides. If a mismatch is detected, a 32 to 52 exonuclease activity is activated and the incorrect base removed. In most organisms DNA polymerases function in a large complex called the replisome that contains multiple accessory subunits, such as the DNA clamp or helicases.

RNA-dependent DNA polymerases are a specialised class of polymerases that copy the sequence of an RNA strand into DNA. They include reverse transcriptase, which is a viral enzyme involved in the infection of cells by retroviruses, and telomerase, which is required for the replication of telomeres. Telomerase is an unusual polymerase because it contains its own RNA template as part of its structure.

Transcription is carried out by a DNA-dependent RNA polymerase that copies the sequence of a DNA strand into RNA. To begin transcribing a gene, the RNA polymerase binds to a sequence of DNA called a promoter and separates the DNA strands. It then copies the gene sequence into a messenger RNA transcript until it reaches a region of DNA called the terminator, where it halts and detaches from the DNA. As with human DNA-dependent DNA polymerases, RNA polymerase II, the enzyme that transcribes most of the genes in the human genome, operates as part of a large protein complex with multiple regulatory and accessory subunits.

Genetic Recombination

A DNA helix usually does not interact with other segments of DNA, and in human cells the different chromosomes even occupy separate areas in the nucleus called "chromosome territories". This physical separation of different chromosomes is important for the ability of DNA to function as a stable repository for information, as one of the few times chromosomes interact is during chromosomal crossover when they recombine. Chromosomal crossover is when two DNA helices break, swap a section and then rejoin.

Recombination allows chromosomes to exchange genetic information and produces new combinations of genes, which increases the efficiency of natural selection and can be important in the rapid evolution of new proteins. Genetic recombination can also be involved in DNA repair, particularly in the cell's response to double-strand breaks.

The most common form of chromosomal crossover is homologous recombination, where the two chromosomes involved share very similar sequences. Non-homologous recombination can be damaging to cells, as it can produce chromosomal translocations and genetic abnormalities. The recombination reaction is catalyzed by enzymes known as *recombinases*, such as RAD51. The first step in recombination is a double-stranded break either caused by an endonuclease or damage to the DNA. A series of steps catalyzed in part by the recombinase then leads to joining of the two helices by at least one Holliday junction, in which a segment of a single strand in each helix is annealed to the complementary strand in the other helix. The Holliday junction is a tetrahedral junction structure that can be moved along the pair of chromosomes, swapping one strand for another. The recombination reaction is then halted by cleavage of the junction and re-ligation of the released DNA.

Evolution of DNA Metabolism

DNA contains the genetic information that allows all modern living things to function, grow and reproduce. However, it is unclear how long in the 4-billion-year history of life DNA has performed this function, as it has been proposed that the earliest forms of life may have used RNA as their genetic material. RNA may have acted as the central part of early cell metabolism as it can both transmit genetic information and carry out catalysis as part of ribozymes. This ancient RNA world where nucleic acid would have been used for both catalysis and genetics may have influenced the evolution of the current genetic code based on four nucleotide bases. This would occur since the number of unique bases in such an organism is a trade-off between a small number of bases increasing replication accuracy and a large number of bases increasing the catalytic efficiency of ribozymes.

Unfortunately, there is no direct evidence of ancient genetic systems, as recovery of DNA from most fossils is impossible. This is because DNA will survive in the environment for less than one million years and slowly degrades into short fragments in solution. Although claims for older DNA have been made, most notably a report of the isolation of a viable bacterium from a salt crystal 250-million years old, these claims are controversial and have been disputed.

Uses in Technology

Genetic Engineering

Modern biology and biochemistry make intensive use of recombinant DNA technology. Recombinant DNA is a man-made DNA sequence that has been assembled from other DNA sequences. They can be transformed into organisms in the form of plasmids or in the appropriate format, by using a viral vector. The genetically modified organisms produced can be used to produce products such as recombinant proteins, used in medical research, or be grown in agriculture.

Forensics

Forensic scientists can use DNA in blood, semen, skin, saliva or hair at a crime scene to identify a perpetrator. This process is called genetic fingerprinting, or more accurately, DNA profiling. In DNA profiling, the lengths of variable sections of repetitive DNA, such as short tandem repeats and minisatellites, are compared between people. This method is usually an extremely reliable technique for identifying a criminal. However, identification can be complicated if the scene is contaminated with DNA from several people. DNA profiling was developed in 1984 by British geneticist Sir Alec Jeffreys, and first used in forensic science to convict Colin Pitchfork in the 1988 Enderby murders case. People convicted of certain types of crimes may be required to provide a sample of DNA for a database. This has helped investigators solve old cases where only a DNA sample was obtained from the scene. DNA profiling can also be used to identify victims of mass casualty incidents.

Bioinformatics

Bioinformatics involves the manipulation, searching, and data mining of DNA sequence data. The development of techniques to store and search DNA sequences have led to widely-applied advances in computer science, especially string searching algorithms, machine learning and database theory. String searching or matching algorithms, which find an occurrence of a sequence of letters inside a larger sequence of letters, were developed to search for specific sequences of nucleotides. In other applications such as text editors, even simple algorithms for this problem usually suffice, but DNA sequences cause these algorithms to exhibit near-worst-case behaviour due to their small number of distinct characters. The related problem of sequence alignment aims to identify homologous sequences and locate the specific mutations that make them distinct. These techniques, especially multiple sequence alignment, are used in studying phylogenetic relationships and protein function. Data sets representing entire genomes' worth of DNA sequences, such as those produced by the Human Genome Project, are difficult to use without annotations, which label the locations of genes and regulatory elements on each chromosome. Regions of DNA sequence that have the characteristic patterns associated with protein- or RNA-coding genes can be identified by gene finding algorithms, which allow researchers to predict the presence of particular gene products in an organism even before they have been isolated experimentally.

DNA Nanotechnology

DNA nanotechnology uses the unique molecular recognition properties of DNA and other nucleic acids to create self-assembing branched DNA complexes with useful properties. DNA is thus used as a structural material rather than as a carrier of biological information. This has lead to the creation of two-dimensional periodic lattices (both tile-based as well as using the "DNA origami" method) as well as three-dimensional structures in the shapes of polyhedra. Nanomechanical devices and algorithmic self-assembly have also been demonstrated, and these DNA structures have been used to template the arrangement of other molecules such as gold nanoparticles and streptavidin proteins.

DNA and Computation

DNA was first used in computing to solve a small version of the directed Hamiltonian path problem, an NP-complete problem DNA computing is advantageous over electronic computers in power use, space use, and efficiency, due to its ability to compute in a highly parallel fashion. A number of other problems, including simulation of various abstract machines, the boolean satisfiability problem, and the bounded version of the travelling salesman problem, have since been analysed using DNA computing. Due to its compactness, DNA also has a theoretical role in cryptography, where in particular it allows unbreakable one-time pads to be efficiently constructed and used.

History and Anthropology

Because DNA collects mutations over time, which are then inherited, it contains historical information and by comparing DNA sequences, geneticists can infer the evolutionary history of organisms, their phylogeny. This field of phylogenetics is a powerful tool in evolutionary biology. If DNA sequences within a species are compared, population geneticists can learn the history of particular populations. This can be used in studies ranging from ecological genetics to anthropology; for example, DNA evidence is being used to try to identify the Ten Lost Tribes of Israel.

DNA has also been used to look at modern family relationships, such as establishing family relationships between the descendants of Sally Hemings and Thomas Jefferson. This usage is closely related to the use of DNA in criminal investigations detailed above. Indeed, some criminal investigations have been solved when DNA from crime scenes has matched relatives of the guilty individual.

History

DNA was first isolated by the Swiss physician Friedrich Miescher who, in 1869, discovered a microscopic substance in the pus of discarded surgical bandages. As it resided in the nuclei of cells, he called it "nuclein". In 1919 this discovery was followed by Phoebus Levene's identification of the base, sugar and phosphate nucleotide unit. Levene suggested that DNA consisted of a string of nucleotide units linked together through the phosphate groups. However, Levene thought the chain was short and the bases repeated in a fixed order. In 1937 William Astbury produced the first X-ray diffraction patterns that showed that DNA had a regular structure.

In 1943, Oswald Theodore Avery discovered that traits of the "smooth" form of the *Pneumococcus* could be transferred to the "rough" form of the same bacteria by mixing killed "smooth" bacteria with the live "rough" form. Avery, along with coworkers Colin MacLeod and Maclyn McCarty, identified DNA as this transforming principle. DNA's role in heredity was confirmed in 1953, when Alfred Hershey and Martha Chase in the Hershey-Chase experiment showed that DNA is the genetic material of the T2 phage.

In 1953, based on X-ray diffraction images taken by Rosalind Franklin and the information that the bases were paired, James D. Watson and Francis Crick suggested what is now accepted as the first accurate model of DNA structure in the journal *Nature*. Experimental evidence for Watson and Crick's model were published in a series of five articles in the same issue of *Nature*. Of these, Franklin and Raymond Gosling's paper was the first publication of X-ray diffraction data that supported the Watson and Crick model, this issue also contained an article on DNA structure by Maurice Wilkins and his colleagues. In 1962, after Franklin's death, Watson, Crick, and Wilkins jointly received the Nobel Prize in Physiology or Medicine. However, speculation continues on who should have received credit for the discovery, as it was based on Franklin's data.

In an influential presentation in 1957, Crick laid out the "Central Dogma" of molecular biology, which foretold the relationship between DNA, RNA, and proteins, and articulated the "adaptor hypothesis". Final confirmation of the replication mechanism that was implied by the double-helical structure followed in 1958 through the Meselson-Stahl experiment. Further work by Crick and coworkers showed that the genetic code was based on non-overlapping triplets of bases, called codons, allowing Har Gobind Khorana, Robert W. Holley and Marshall Warren Nirenberg to decipher the genetic code. These findings represent the birth of molecular biology.

DYNAMICAL GENETICS

Dynamical genetics concerns the study and the interpretation of those phenomena in which physiological enzymatic protein complexes alter the DNA, in a more or less sophisticated way.

The study of such mechanisms is important firstly since they promote useful functions, as for example the immune system recombination (on individual scale) and the crossing-over (on evolutionary scale); secondly since they may sometimes become harmful because of some malfunctioning, causing for example neurodegenerative disorders.

Typical examples of dynamical genetics subjects are:

- dynamic mutations, term introduced by Robert I. Richards and Grant R. Sutherland to indicate mutations caused by other mutations; this phenomenon often involves the Variable Number Tandem Repeats, closely related to many neurodegenerative diseases, as the trinucleotide repeat disorders (interpreted by Anita Harding).
- dynamic genome, term introduced by Nina Fedoroff and David Botstein to indicate the transposition discovered by Barbara McClintock.
- immune V(D)J recombination (discovered by Susumu Tonegawa) and isotype class switching, terms introduced to indicate two kinds of immune system recombinations, which are the main cause of the enormous variety of antibodies.
- horizontal DNA transfer (discovered by Frederick Griffith) that indicates the DNA transfer between two organisms.
- crossing-over (discovered by Thomas Hunt Morgan) mediated by formation and unwinding (by means of peculiar enzymatic complexes such as helicase) of uncommon four-helix DNA structures known as G-quadruplexes (discovered by Martin Gellert, Marie N. Lipsett, and David R. Davies).

HUMAN CLONING

Although genes are recognized as influencing behavior and cognition, "genetically identical" does *not* mean altogether identical; identical twins, despite being natural human clones with near identical DNA, are separate people, with separate experiences and not altogether overlapping personalities. The relationship between an "original" and a clone is rather like that between identical triplets raised apart; they share nearly all of the same DNA, but little of the same environment. A lively scientific debate on this topic occurred in the journal *Nature* in 1997. Ultimately, the question of how similar an original and a clone would be boils down to how much of personality is determined by genetics, an area still under active scientific investigation.

Although genes are recognized as influencing behavior and cognition, "genetically identical" does *not* mean altogether identical; identical twins, despite being natural human clones with near identical DNA, are separate people, with separate experiences and not altogether overlapping personalities. The relationship between an "original" and a clone is rather like that between identical triplets raised apart; they share nearly all of the same DNA, but little of the same environment. A lively scientific debate on this topic occurred in the journal *Nature* in 1997. Ultimately, the question of how similar an original and a clone would be boils down to how much of personality is determined by genetics, an area still under active scientific investigation.

Claims of Success in Human Cloning Beyond the Embryo Stage

In 1978 David Rorvik claimed in his book *In His Image: The Cloning of a Man* that he had personal knowledge of the creation of a human clone. A court case followed. He failed to produce corroborating evidence to back up his claims; now regarded as a hoax.

Severino Antinori made claims in November, 2002 that a project to clone human beings had succeeded, with the first human clone due to be born [in January 2003.] His claims were received with skepticism from many observers.

In December 2002, Clonaid, the medical arm of a religion called Raëlism, who believe that aliens introduced human life on Earth, claimed to have successfully cloned a human being. They claim that aliens taught them how to perform cloning, even though the company has no record of having successfully cloned any previous animal. A spokesperson said an independent agency would prove that the baby, named Evá, is in fact an exact copy of her mother. Shortly thereafter, the testing was cancelled, with the spokesperson claiming the decision would ultimately be left up to Evá's parents.

On October 9, 2003, newspaper *Le journal de Montréal* published an article accusing Clonaid and the Raelian religion of maintaining an outright hoax in its claims regarding cloning a human baby.

In December 2004 Dr. Brigitte Boisselier, claimed in a letter to the UN that Clonaid has successfully cloned 13 children, however their identities cannot be revealed to the public in order to protect them.

In 2004 a group of scientists led by Hwang Woo-Suk of Seoul National University in South Korea claimed to have grown 30 cloned human embryos to the one-week stage, and then successfully harvested stem cells from them. The results of their experiment were published in the peer-reviewed journal *Science*.

On May 30, 2005, Hwang's team announced the creation of 11 lines of human stem cells, using a different technique (Hwang et al. 2005). The journal Science later retracted Hwang's publications when investigations into the matter revealed that the claims were fraudulent.

Possible Advantages

Human cloning might produce many benefits. Human therapeutic cloning could provide genetically identical cells for regenerative medicine, and tissues and organs for transplantation. Such cells, tissues, and organs would neither trigger an immune response nor require the use of immunosuppressive drugs. Both basic research and therapeutic development for serious diseases such as cancer, heart disease, and diabetes, as well as improvements in burn treatment and reconstructive and cosmetic surgery, are areas that might benefit from such new technology.

Human reproductive cloning also might produce benefits. Antinori and Zavos hope to create a fertility treatment that allows parents who are both infertile to have children with at least some of their DNA in their offspring.

Some scientists, including Dr. Richard Seed, suggest that human cloning might obviate the human aging process. How this might work is not entirely clear since the brain or identity would have to be transferred to a cloned body. Dr. Preston Estep has suggested the terms "replacement cloning" to describe the generation of a clone of a previously living person, and "persistence cloning" to describe the production of a cloned body for the purpose of obviating aging, although he maintains that such procedures currently should be considered science fiction.

In Aubrey de Gray's proposed SENS (Strategy for Engineered negligible senescence) one of the considered options to repair the cell depletion related to cellular senescence is to grow replacement tissues from stem cells harvested from a cloned embryo.

Human cloning might produce many benefits. Human therapeutic cloning could provide genetically identical cells for regenerative medicine, and tissues and organs for transplantation. Such cells, tissues, and organs would neither trigger an immune response nor require the use of immunosuppressive drugs. Both basic research and therapeutic development for serious diseases such as cancer, heart disease, and diabetes, as well as improvements in burn treatment and reconstructive and cosmetic surgery, are areas that might benefit from such new technology.

Human reproductive cloning also might produce benefits. Antinori and Zavos hope to create a fertility treatment that allows parents who are both infertile to have children with at least some of their DNA in their offspring.

Some scientists, including Dr. Richard Seed, suggest that human cloning might obviate the human aging process. How this might work is not entirely clear since the brain or identity would have to be transferred to a cloned body. Dr. Preston Estep has suggested the terms "replacement cloning" to describe the generation of a clone of a previously living person, and "persistence cloning" to describe the production of a cloned body for the purpose of obviating aging, although he maintains that such procedures currently should be considered science fiction.

In Aubrey de Gray's proposed SENS (Strategy for Engineered negligible senescence) one of the considered options to repair the cell depletion related to cellular senescence is to grow replacement tissues from stem cells harvested from a cloned embryo.

The European Convention on Human Rights and Biomedicine prohibits human cloning in one of its additional protocols, but this protocol has been ratified only by Greece, Spain and Portugal. The Charter of Fundamental Rights of the European Union explicitly prohibits reproductive human cloning, though the Charter currently carries no legal standing. The proposed Treaty of Lisbon would, if ratified, make the charter legally binding for the institutions of the European Union.

U.S.

In 1998, 2001, and 2003 the U.S. House of Representatives voted whether to ban all human cloning, both reproductive and therapeutic. Each time, divisions in the Senate over therapeutic cloning prevented either competing proposal (a ban on both forms or reproductive cloning only) from passing. Some American states ban both forms of cloning, while some others outlaw only reproductive cloning.

Current regulations prohibit federal funding for research into human cloning, which effectively prevents such research from occurring in public institutions and private institutions such as universities which receive federal funding. However, there are currently no federal laws in the United States which ban cloning completely, and any such laws would raise difficult Constitutional questions similar to the issues raised by abortion.

U.K.

The British government introduced legislation in order to allow licensed therapeutic cloning in a debate in January 2001 after an amendment to the Human Fertilisation & Embryology Act 1990. However on November 15, 2001 a prolife group won a High Court legal challenge that effectively left cloning unregulated in the UK. Their hope was that Parliament would fill this gap by passing prohibitive legislation. The government was quick to pass legislation prohibiting reproductive cloning Human Reproductive Cloning Act 2001. The remaining gap with regard to therapeutic cloning was closed when the appeals courts reversed the previous decision of the High Court. Currently therapeutic cloning is allowed under license from the Human Fertilisation and Embryology Authority. The first licence was granted on August 11,

2004 to researchers at the University of Newcastle to allow them to investigate treatments for diabetes, Parkinson's disease and Alzheimer's disease.

GENE NOMENCLATURE

Gene nomenclature is the scientific naming of genes, the units of heredity in living organisms. An international committee published recommendations for genetic symbols and nomenclature in 1957. The need to develop formal guidelines for human gene names and symbols was recognized in the 1960's and full guidelines were issued in 1979 (Edinburgh Human Genome Meeting). Several other species-specific research communities (e.g., Drosophila, mouse) have adopted nomenclature standards, as well, and have published them on the relevant model organism websites and in scientific journals, including the *Trends in Genetics* Genetic Nomenclature Guide. Scientists familiar with a particular gene family may work together to revise the nomenclature for the entire set of genes when new information becomes available. For many genes and their corresponding proteins, however, an assortment of alternate names is in use across the scientific literature and public biological databases, thus posing a challenge to effective organization and exchange of biological information.

Nomenclature Guidelines

Species-specific resources

The HUGO Gene Nomenclature Committee (HGNC) is responsible for providing human gene naming guidelines and approving new, unique human gene names and symbols (short form abbreviations). For some non-human species, model organism databases serve as central repositories of guidelines and help resources, including advice from curators and nomenclature committees. In addition to species-specific databases, approved gene names and symbols for many species can be located in the National Center for Biotechnology Information's Entrez Gene database.

Invertebrates

- Fly (*Drosophila melanogaster*): Genetic nomenclature for *Drosophila melanogaster* at Flybase.
- Worm (*Caenorhabditis elegans*): Genetic Nomenclature for *Caenorhabditis elegans*.

Plants

- Maize (*Zea mays*): A Standard For Maize Genetics Nomenclature at MaizeGDB.
- Thale cress (*Arabidopsis thaliana*): Arabidopsis Nomenclature at The Arabidopsis Information Resource (TAIR).

Slime Molds

- Dictyostelid (*Dictyostelium discoideum*): Nomenclature Guidelines at dictyBase.

Vertebrates

- Human (*Homo sapiens*): Guidelines for Human Gene Nomenclature.
- Mouse (*Mus musculus*) and rat (*Rattus norvegicus*): Rules for Nomenclature of Genes, Genetic Markers, Alleles, and Mutations in Mouse and Rat at Mouse Genome Informatics (MGI).
- Frog (*Xenopus laevis* and *tropicalis*): Suggested Xenopus Gene Name Guidelines at Xenbase.
- Zebrafish (*Danio rerio*): Zebrafish Nomenclature Guidelines at the Zebrafish Model Organism Database (ZFIN).

Yeast

- Budding yeast (*Saccharomyces cerevisiae*): SGD Gene Naming Guidelines at *Saccharomyces* Genome Database.
- Candida (*Candida albicans*): *C. albicans* Gene Nomenclature Guide at *Candida* Genome Database (CGD).
- Fission yeast (*Schizosaccharomyces pombe*): Gene Naming Guidelines.

Vertebrate Gene Symbol Formatting

The research communities of vertebrate model organisms have adopted guidelines whereby genes in these species are given, whenever possible, the same names as their human orthologs. The use of prefixes on gene symbols to indicate species (e.g., "Z" for zebrafish) is discouraged. The recommended formatting of printed gene and protein symbols varies between species.

Human

Gene symbols generally are italicised, with all letters in uppercase (e.g., *SHH*, for sonic hedgehog). Italics are not necessary in gene catalogs. Protein designations are the same as the gene symbol, but are not italicised; all letters are in uppercase (SHH). mRNAs and cDNAs use the same formatting conventions as the gene symbol.

Mouse and Rat

Gene symbols generally are italicised, with only the first letter in uppercase and the remaining letters in lowercase (*Shh*). Italics are not required on web pages. Protein designations are the same as the gene symbol, but are not italicised; all letters are in uppercase (SHH).

Frog (Xenopus *sp.)*

Gene symbols are not italicised and all letters are in lowercase (shh). Protein designations are the same as the gene symbol, are not italicised, and all letters are in uppercase (SHH).

Zebrafish

Gene symbols are italicised, with all letters in lowercase (*shh*). Protein designations are the same as the gene symbol, but are not italicised; the first letter is in uppercase and the remaining letters are in lowercase (Shh).

CHROMOSOME

Chromosomes are the organized form of DNA found in cells. Chromosomes contain a single continuous piece of DNA, which contains many genes, regulatory elements and other intervening nucleotide sequences. A broader definition of "chromosome" also includes the DNA-bound proteins which serve to package and manage the DNA. The word *chromosome* comes from the Greek ÷ñöìá (*chroma*, color) and óöìá (*soma*, body) due to their property of being stained very strongly with vital and supravital dyes.

Chromosomes vary extensively between different organisms. The DNA molecule may be circular or linear, and can contain anything from tens of kilobase pairs to hundreds of megabase pairs. Typically eukaryotic cells (cells with nuclei) have large linear chromosomes and prokaryotic cells (cells without nuclei) smaller circular chromosomes, although there are many exceptions to this rule. Furthermore, cells may contain more than one type of chromosome; for example mitochondria in most eukaryotes and chloroplasts in plants have their own small chromosome in addition to the nuclear chromosomes.

In eukaryotes, nuclear chromosomes are packaged by proteins into a condensed structure called chromatin. This allows the massively-long DNA molecules to fit into the cell nucleus. The structure of chromatin varies through the cell cycle, and is responsible for the organisation of chromosomes into the classic four-arm structure during mitosis and meiosis.

"Chromosome" is a rather loosely defined term. In prokaryotes, a small circular DNA molecule may be called either a plasmid or a small chromosome. These small circular genomes are also found in mitochondria, and chloroplasts, reflecting their bacterial origins. The simplest chromosomes are found in viruses: these DNA or RNA molecules are short linear or circular chromosomes that often lack any structural proteins.

History

Chromosomes were first observed in plant cells by a Swiss botanist named Karl Wilhelm von Nägeli in 1842, and independently in Ascaris worms by Belgian scientist Edouard Van Beneden (1846-1910). The use of basophilic aniline dyes was a fundamentally new technique for effectively staining the chromatin material in the nucleus. Their behavior in animal (salamander) cells was later described in detail by German cytologist and professor of anatomy Walther Flemming, the discoverer of mitosis, in 1882. The name was invented later by another German anatomist, Heinrich von Waldeyer in 1888.

Chromosomes in Eukaryotes

Eukaryotes (cells with nuclei such as plants, yeast, and animals) possess multiple large linear chromosomes contained in the cell's nucleus. Each chromosome has one centromere, with one or two arms projecting from the centromere, although under most circumstances these arms are not visible as such. In addition most eukaryotes have a small circular mitochondrial genome, and some eukaryotes may have additional small circular or linear cytoplasmic chromosomes.

In the nuclear chromosomes of eukaryotes, the uncondensed DNA exists in a semi-ordered structure, where it is wrapped around histones (structural proteins), forming a composite material called chromatin.

Chromatin

Chromatin is the complex of DNA and protein found in the eukaryotic nucleus which packages chromosomes. The structure of chromatin varies significantly between different stages of the cell cycle, according to the requirements of the DNA.

Interphase Chromatin

During interphase (the period of the cell cycle where the cell is not dividing) two types of chromatin can be distinguished:

- Euchromatin, which consists of DNA that is active, e.g., expressed as protein.
- Heterochromatin, which consists of mostly inactive DNA. It seems to serve structural purposes during the chromosomal stages. Heterochromatin can be further distinguished into two types:
  - *Constitutive heterochromatin*, which is never expressed. It is located around the centromere and usually contains repetitive sequences.
  - *Facultative heterochromatin*, which is sometimes expressed.

Individual chromosomes cannot be distinguished at this stage - they appear in the nucleus as a homogeneous tangled mix of DNA and protein.

Metaphase Chromatin and Division

In the early stages of mitosis or meiosis (cell division), the chromatin strands become more and more condensed. They cease to function as accessible genetic material (transcription stops) and become a compact transportable form. This compact form makes the individual chromosomes visible, and they form the classic four arm structure, a pair of sister chromatids attached to each other at the centromere. The shorter arms are called *p arms* (from the French *petit*, small) and the longer arms are called *q arms* (*q* follows *p* in the Latin alphabet). This is the only natural context in which individual chromosomes are visible with an optical microscope.

During divisions long microtubules attach to the centromere and the two opposite ends of the cell. The microtubules then pull the chromatids apart, so that each daughter cell inherits one set of chromatids. Once the cells have divided, the chromatids are uncoiled and can function again as chromatin. In spite of their appearance, chromosomes are structurally highly condensed which enables these giant DNA structures to be contained within a cell nucleus.

The self assembled microtubules form the spindle, which attaches to chromosomes at specialized structures called kinetochores, one of which is present on each sister chromatid. A special DNA base sequence in the region of the kinetochores provides, along with special proteins, longer-lasting attachment in this region.

Chromosomes in Prokaryotes

Prokaryotes (eg. Bacteria) typically have a single circular chromosome, but many variations do exist. Bacterial DNA also exists as plasmids, essentially miniature chromosomes, which are small circular pieces of DNA that are readily transmitted between bacteria. The distinction between plasmids and chromosomes is poorly defined, though size and necessity are generally taken into account.

Structure in Sequences

Prokaryotes chromosomes have less sequence-based structure than eukaryotes. They do, however, typically have a single point, the origin of replication, from which replication starts.

The genes in prokaryotes are often organised in operons, and do not contain introns, unlike eukaryotes.

Location in the Cell

Bacterial chromosomes tend to be tethered to the plasma membrane of the bacteria. In molecular biology application, this allows for its isolation from plasmid DNA by centrifugation of lysed bacteria and pelleting of the membranes (and the attached DNA).

DNA Packaging

Prokaryotes do not possess histones or nuclei, and so do not possess chromatin like eukaryotes. There is, however, thought to be some structural organisation to help condense the large molecule into the small prokaryotic cell.

Prokaryotic chromosomes and plasmids are, like eukaryotic DNA, generally supercoiled. The DNA must first be released into its relaxed state for access for transcription, regulation, and replication.

Number of Chromosomes in Various Organisms

Eukaryotes

Chromosome numbers (2n) in some plants

Plant Species	#
Arabidopsis thaliana	10
Rye	14
Maize	20
Einkorn wheat	14
Durum wheat	28
Bread wheat	42
Wild tobacc	24
Cultivated tobacc	48
Adder's Tongue Fern	1262

Chromosome numbers (2n) in some animals

Species	#	*Species*	#
Common fruit fly	8	Guinea Pig	64
Dove	16	Snail	24
Earthworm	36	Tibetan fox	36
Domestic cat	38	Domestic pig	38
Lab mouse	40	Lab rat	42
Rabbit	44	Syrian hamster	44
Hare	46	Human	46

Gorillas, Chimpanzees	48	Domestic sheep	54
Elephants	56	Cow	60
Donkey	62	Horse	64
Dog	78	Chicken	39
Goldfish	100-104	Silkworm	28

Normal members of a particular eukaryotic species all have the same number of nuclear chromosomes. Other eukaryotic chromosomes, i.e., mitochondrial and plasmid-like small chromosomes, are much more variable in number, and there may be thousands of copies per cell.

Asexually reproducing species have one set of chromosomes, which is the same in all body cells.

Sexually reproducing species have somatic cells (body cells), which are diploid [2n] having two sets of chromosomes, one from the mother and one from the father. Gametes, reproductive cells, are haploid [n]: they have one set of chromosomes. Gametes are produced by meiosis of a diploid germ line cell. During meiosis, the matching chromosomes of father and mother can exchange small parts of themselves (crossover), and thus create new chromosomes that are not inherited solely from either parent. When a male and a female gamete merge (fertilization), a new diploid organism is formed.

Some animal and plant species are polyploid [Xn]: they have more than two sets of homologous chromosomes. Agriculturally important plants such as tobacco or wheat are often polyploid compared to their ancestral species. Wheat has a haploid number of seven chromosomes, still seen in some cultivars as well as the wild progenitors. The more common pasta and bread wheats are polyploid, having 28 (tetraploid) and 42 (hexaploid) chromosomes compared to the 14 (diploid) chromosomes in the wild wheat.

Historical note: In 1921, Theophilus Painter claimed, based on his observations, that human sex cells had 24 chromosomes each, giving humans 48 chromosomes total. It wasn't until 1955 that the number of chromosomes was clearly shown to be 23.

Prokaryotes

Prokaryote species generally have one copy of each major chromosome, but most cells can easily survive with multiple copies. Plasmids and plasmid-like small chromosomes are, like in eukaryotes, very variable in copy number. The number of plasmids in the cell is almost entirely determined by the rate of division of the plasmid - fast division causes high copy number, and vice versa.

Karyotype

Karyotyping is a technique used to determine the (diploid) number of nuclear chromosomes of a eukaryotic organism, and may be used for determining sex and spotting chromosomal abnormalities. Cells can be locked part way through division (in metaphase) in vitro (in a reaction vial) with colchicine. These cells are then stained, photographed and arranged into a karyotype, also called *karyogram*.

Like many sexually reproducing species, humans have special gonosomes (sex chromosomes, in contrast to autosomes). These are XX in females and XY in males, and can be seen in the karyotype.

Chromosomal Aberrations

Chromosomal aberrations are disruptions in the normal chromosomal content of a cell, and are a major cause of genetic conditions in humans, such as Down syndrome. Some chromosome abnormalities do not cause disease in carriers, such as translocations, or chromosomal inversions, although they may lead to a higher chance of having a child with a chromosome disorder. Abnormal numbers of chromosomes or chromosome sets, aneuploidy, may be lethal or give rise to genetic disorders. Genetic counseling is offered for families that may carry a chromosome rearrangement.

The gain or loss of chromosome material can lead to a variety of genetic disorders. Human examples include:

Cri du chat, which is caused by the deletion of part of the short arm of chromosome 5. "Cri du chat" means "cry of the cat" in French, and the condition was so-named because affected babies make high-pitched cries that sound like a cat. Affected individuals have wide-set eyes, a small head and jaw and are moderately to severely mentally retarded and very short.

Wolf-Hirschhorn syndrome, which is caused by partial deletion of the short arm of chromosome 4. It is characterized by severe growth retardation and severe to profound mental retardation.

Down's syndrome, usually is caused by an extra copy of chromosome 21 (trisomy 21). Characteristics include decreased muscle tone, asymmetrical skull, slanting eyes and mild to moderate mental retardation.

Edwards syndrome, which is the second most common trisomy after Down syndrome. It is a trisomy of chromosome 18. Symptoms include mental and motor retardation and numerous congenital anomalies causing serious health problems. Ninety percent die in infancy; however, those who live past their first birthday usually are quite healthy thereafter. They have a characteristic hand appearance with clenched hands and overlapping fingers.

Patau Syndrome, also called D-Syndrome or trisomy-13. Symptoms are somewhat similar to those of trisomy-18, but they do not have the characteristic hand shape.

Idic15, abbreviation for Isodicentric 15 on chromosome 15; also called the following names due to various researches, but they all mean the same; IDIC(15), Inverted dupliction 15, extra Marker, Inv dup 15, partial tetrasomy 15

Jacobsen syndrome, also called the terminal 11q deletion disorder. This is a very rare disorder. Those affected have normal intelligence or mild mental retardation, with poor expressive language skills. Most have a bleeding disorder called Paris-Trousseau syndrome.

Klinefelter's syndrome (XXY). Men with Klinefelter syndrome are usually sterile, and tend to have longer arms and legs and to be taller than their peers. Boys with the syndrome are often shy and quiet, and have a higher incidence of speech delay and dyslexia. During puberty, without testosterone treatment, some of them may develop gynecomastia.

Turner syndrome (X instead of XX or XY). In Turner syndrome, female sexual characteristics are present but underdeveloped. People with Turner syndrome often have a short stature, low hairline, abnormal eye features and bone development and a "caved-in" appearance to the chest.

XYY syndrome. XYY boys are usually taller than their siblings. Like XXY boys and XXX girls, they are somewhat more likely to have learning difficulties.

Triple-X syndrome (XXX). XXX girls tend to be tall and thin. They have a higher incidence of dyslexia.

Small supernumerary marker chromosome. This means there is an extra, abnormal chromosome. Features depend on the origin of the extra genetic material. Cat-eye syndrome and isodicentric chromosome 15 syndrome (or Idic15) are both caused by a supernumerary marker chromosome, as is Pallister-Killian syndrome. Chromosomal mutations produce changes in whole chromosomes (more than one gene) or in the number of chromosomes present.

Human Chromosomes

Human cells have 23 pairs of large linear nuclear chromosomes, giving a total of 46 per cell. In addition to these, human cells have many hundreds of copies of the mitochondrial genome. Sequencing of the human genome has provided a great deal of information about each of the chromosomes. Below is a table compiling statistics for the chromosomes, based on the Sanger Institute's human genome information in the Vertebrate Genome Annotation (VEGA) database. Number of genes is an estimate as it is in part based on gene predictions. Total chromosome length is an estimate as well, based on the estimated size of unsequenced heterochromatin regions.

Chromosome	*Genes*	*Total bases*	*Sequenced bases*
1	3,148	247,200,000	224,999,719
2	902	242,750,000	237,712,649
3	1,436	199,450,000	194,704,827
4	453	191,260,000	187,297,063
5	609	180,840,000	177,702,766
6	1,585	170,900,000	167,273,992
7	1,824	158,820,000	154,952,424
8	781	146,270,000	142,612,826
9	1,229	140,440,000	120,312,298
10	1,312	135,370,000	131,624,737
11	405	134,450,000	131,130,853
12	1,330	132,290,000	130,303,534
13	623	114,130,000	95,559,980
14	886	106,360,000	88,290,585
15	676	100,340,000	81,341,915
16	898	88,820,000	78,884,754
17	1,367	78,650,000	77,800,220
18	365	76,120,000	74,656,155
19	1,553	63,810,000	55,785,651
20	816	62,440,000	59,505,254
21	446	46,940,000	34,171,998
22	595	49,530,000	34,893,953
X (sex chromosome)	1,093	154,910,000	151,058,754
Y (sex chromosome)	125	57,740,000	22,429,293

ASEXUAL REPRODUCTION

Asexual reproduction is a form of reproduction which does not involve meiosis, ploidy reduction, or fertilization. Asexual reproduction only takes one parent. A more stringent definition is *agamogenesis* which refers to reproduction without the fusion of gametes. Asexual reproduction is the primary form of reproduction for single-celled organisms such as archaea, bacteria, and protists. Many plants and fungi reproduce mostly asexually as well. While all prokaryotes reproduce asexually (without the formation and fusion of gametes), mechanisms for lateral gene transfer such as conjugation, transformation and transduction are sometimes likened to sexual reproduction.

Asexual reproduction is relatively rare among multicellular organisms, for reasons that are not completely understood. Current hypotheses suggest that, while asexual reproduction may have short term benefits when rapid population growth is important or in stable environments, sexual reproduction offers a net advantage by allowing more rapid generation of genetic diversity, allowing adaptation to changing environments.

Because asexual reproduction does not require the formation of gametes (often in separate individuals) and bringing them together for fertilization, it occurs much faster than sexual reproduction and requires less energy. Asexual lineages can increase their numbers rapidly because all members can reproduce viable offspring. In sexual populations with two genders, some of the individuals are male and cannot themselves produce offspring. This means that an asexual lineage will have roughly double the rate of population growth under ideal conditions when compared with a sexual population half composed of males. This is known as the two-fold cost of sex. Other advantages include the ability to reproduce without a partner in situations where the population density is low (such as for some desert lizards), reducing the chance of finding a mate, or during colonisation of isolated habitats such as oceanic islands, where a single (female) member of the species is enough to start a population.

Another consequence of asexual reproduction, which may have both benefits and costs, is that offspring are typically genetically similar to their parent, with as broad a range as that individual receives from one parent. The lack of genetic recombination results in fewer genetic alternatives than with sexual reproduction. Many forms of asexual reproduction, for example budding or fragmentation, produce an exact replica of the parent. This genetic similarity may be beneficial if the genotype is well-suited to a stable environment, but disadvantageous if the environment is changing. For example, if a new predator or pathogen appears and a genotype is particularly defenseless against it, an asexual lineage is more likely to be completely wiped out by it. In contrast, a lineage that reproduces sexually has a higher probability of having more members survive due to the genetic recombination that produces a novel genotype in each individual. Similar arguments apply to changes in the physical environment. From an evolutionary standpoint, one could thus argue that asexual reproduction is inferior because it stifles the potential for change. However, there is also a significantly reduced chance of mutation or other complications that can result from the mixing of genes. A 2004 article in the journal *Nature* reported that the modern arbuscular mycorrhizas fungi, which reproduces asexually, is identical to fossil records dating back to the Ordovician period, 460 million years ago.

Types of Asexual Reproduction

Binary Fission

Many, but not all, single-celled organisms (unicellular), such as archaea, bacteria, and protists, reproduce asexually through binary fission. An exception to the rule are unicellular fungi such as fission yeast, unicellular algae such as *Chlamydomonas*, and ciliates and some other protists, which reproduce both sexually and asexually. Some single-celled organisms(unicellular) rely on one or more host organisms in order to reproduce, but most literally divide into two organisms.

Budding

Some cells split via budding (for example baker's yeast), resulting in a smaller 'mother' cell and larger 'daughter' cell. Offspring is larger than parent. Budding is also known on a multicellular level. An animal example is hydra, which reproduces by budding. The buds grow into fully matured individuals which eventually break away from the parent animal.

Vegetative Reproduction

Vegetative reproduction is a type of asexual reproduction found in plants where new independent individuals are formed without the production of seeds or spores. Examples for vegetative reproduction include the formation of plantlets on specialized leaves (for example in kalanchoe), the growth of new plants out of rhizomes or stolons (for example in strawberry), or the formation of new bulbs (for example in tulips). The resulting plants form a clonal colony.

Spore Formation

Many multicellular organisms form spores during their biological life cycle in a process called *sporogenesis*. Exceptions are animals and some protists, who undergo *gametic meiosis* immediately followed by fertilization. Plants and many algae on the other hand undergo *sporic meiosis* where meiosis leads to the formation of haploid spores rather than gametes. These spores grow into multicellular individuals (called gametophytes in the case of plants) without a fertilization event. This haploid individual gives rise to gametes through mitosis. Meiosis and gamete formation therefore occur in separate generations or "phases" of the life cycle, referred to as alternation of generations. Since sexual reproduction is often more narrowly defined as the fusion of gametes (fertilization), spore formation in plant sporophytes and algae might be considered a form of asexual reproduction (agamogenesis) despite being the result of meiosis and undergoing a reduction in ploidy. However, both events (spore formation and fertilization) are necessary to complete sexual reproduction in the plant life cycle.

Fungi and some algae can also utilize true asexual spore formation, which involves mitosis giving rise to reproductive cells called mitospores that develop into a new organism after dispersal. This method of reproduction is found for example in conidial fungi and the red alga *Polysiphonia*, and involves sporogenesis without meiosis. Thus the chromosome number of the spore cell is the same as that of the parent producing the spores. However, mitotic sporogenesis is an exception and most spores, such as those of plants and most fungi and algae, are produced by meiosis.

Fragmentation

Fragmentation is a form of asexual reproduction where a new organism grows from a fragment of the parent. Each fragment develops into a mature, fully grown individual. Fragmentation is seen in many organisms such as animals (some annelid worms and starfish), fungi, and plants. Some plants have specialized structures for reproduction via fragmentation, such as *gemmae* in liverworts. Most lichens, which are a symbiotic union of a fungus and photosynthetic algae or bacteria, reproduce through fragmentation to ensure that new individuals contain both symbionts. These fragments can take the form of *soredia*, dust-like particles consisting of fungal hyphae wrapped around photobiont cells.

Parthenogenesis

Parthenogenesis is a form of agamogenesis in which an unfertilized egg develops into a new individual. Parthenogenesis occurs naturally in many plants, invertebrates (e.g. water fleas, aphids, some bees and parasitic wasps), and vertebrates (e.g. some reptiles, amphibians, fish, and, very rarely birds). In plants, apomixis may or may not involve parthenogenesis.

Agamogenesis

Agamogenesis is any form of reproduction that does not involve a male gamete. Examples are parthenogenesis and apomixis.

Apomixis and Nucellar Embryony

Apomixis in plants is the formation of a new sporophyte without fertilization. It is important in ferns and in flowering plants, but is very rare in other seed plants. In flowering plants, the term "apomixis" is now most often used for agamospermy, the formation of seeds without fertilization, but was once used to include vegetative reproduction. An example of an apomictic plant would be the triploid European dandelion. Apomixis mainly occurs in two forms: In gametophytic apomixis, the embryo arises from an unfertilized egg within a diploid embryo sac that was formed without completing meiosis. In nucellar embryony, the embryo is formed from the diploid nucellus tissue surrounding the embryo sac. Nucellar embryony occurs in some citrus seeds. Male apomixis can occur in rare cases, such as the Saharan Cypress where the genetic material of the embryo are derived entirely from pollen. The term "apomixis" is also used for asexual reproduction in some animals, notably water-fleas, Daphnia.

Alternation Between Sexual and Asexual Reproduction

Some species alternate between the sexual and asexual strategies, an ability known as heterogamy, depending on conditions. For example, the freshwater crustacean *Daphnia* reproduces by parthenogenesis in the spring to rapidly populate ponds, then switches to sexual reproduction as the intensity of competition and predation increases. Many protists and fungi alternate between sexual and asexual reproduction. For example, the slime mold *Dictyostelium* undergoes binary fission as single-celled amboebae under favorable conditions. However, when conditions turn unfavorable, the cells aggregate and switch to sexual reproduction leading to the formation of spores. The hyphae of the common mold (*Rhizopus*) are capable of producing both mitotic as well as meiotic spores. Many algae similarly switch between sexual and asexual reproduction. Asexual reproduction is far less complicated than sexual reproduction. In sexual reproduction one must find a mate.

Examples in Animals

A number of invertebrates and some less advanced vertebrates are known to alternate between sexual and asexual reproduction, or be exclusively asexual. Alternation is observed in a few types of insects, such as aphids (which will, under favourable conditions, produce eggs that have not gone through meiosis, essentially cloning themselves) and the cape bee Apis mellifera capensis (which can reproduce asexually through a process called thelytoky). A few species of amphibians and reptiles have the same ability (see parthenogenesis for concrete examples). A very unusual case among more advanced vertebrates is the female turkey's ability to produce fertile eggs in the absence of a male. The eggs result in often sickly, and nearly always male turkeys. This behaviour can interfere with the incubation of eggs in turkey farming.

SEXUAL REPRODUCTION

Sexual reproduction is a union that results in increasing genetic diversity of the offspring. It is characterized by two processes: meiosis, involving the halving of the number of chromosomes; and fertilization, involving the fusion of two gametes and the restoration of the original number of chromosomes. During meiosis, the chromosomes of each pair usually cross over to achieve genetic recombination. The evolution of sex is a major puzzle. The first fossilized evidence of sexually reproducing organisms is from eukaryotes of the Stenian period, about 1.2 to 1 billion years ago. Sexual reproduction is the primary method of reproduction for the vast majority of visible organisms, including almost all animals and plants. Bacterial conjugation, the transfer of DNA between two bacteria, is often mistakenly confused with sexual reproduction, because the mechanics are similar. A major question is why sexual reproduction persists when parthenogenesis appears in some ways to be a superior form of reproduction. Contemporary evolutionary thought proposes some explanations. It may be due to selection pressure on the clade itself—the ability for a population to radiate more rapidly due to a changing environment through sexual recombination than parthenogenesis allows. Alternatively, sexual reproduction may allow for the 'ratcheting' of evolutionary speed as one clade competes with another for a limited resource.

Reproduction in Plants

In flowering plants, the anther produces male gametophytes, the sperm is produced in pollen grains, which attach to the stigma on top of a carpel, in which the female gametophytes (inside ovules) are located. After the pollen tube grows through the carpel's style, the sperm cell nuclei from the pollen grain migrate into the ovule to fertilize the egg cell and endosperm nuclei within the female gametophyte in a process termed *double fertilization*. The resulting zygote develops into an embryo, while the triploid endosperm (one sperm cell plus two female cells) and female tissues of the ovule give rise to the surrounding tissues in the developing seed. The ovary, which produced the female gametophyte(s), then grows into a fruit, which surrounds the seed(s). Plants may either self-pollinate or cross-pollinate. Nonflowering plants like ferns, moss and liverworts use other mean of sexually reproduction.

Reproduction in Archosaurs (Reptiles and Birds)

Male and female birds and reptiles both have cloacae, an opening through which eggs, sperm, and wastes pass. Intercourse is performed by pressing the lips of the cloacae together, during which time the male transfers his sperm to the female. The female lays amniotic eggs in which the young gestate. Nevertheless, a few species, including most waterfowl and ostriches, have a phallus shaped organ analogous to the mammals' penis.

Reproduction in Mammals

There are three kind of mammals; Monotremes, Placentals and Marsupials, all with internal fertilisation yet do still differ from each other. In placental mammals, offspring are born as juveniles: complete animals with the sex organs present although not reproductively functional. After several months or years, the sex organs develop further to maturity and the animal becomes sexually mature. Most female mammals are only fertile during certain periods and during those times, they are said to be "in heat". At this point, the animal is ready to mate. Individual male and female mammals meet and carry out copulation. For most mammals, males and females exchange sexual partners throughout their adult lives.

The Mammalian Male

The male reproductive system contains two main divisions: the penis, and the testes, the latter of which is where sperm are produced. In humans, both of these organs are outside the abdominal cavity, but they can be primarily housed within the abdomen in other animals (for instance, in dogs, the penis is internal except when mating). Having the testes outside the abdomen best facilitates temperature regulation of the sperm, which require specific temperatures to survive. Sperm are the smaller of the two gametes and are generally very short-lived, requiring males to produce them continuously from the time of sexual maturity until death. Prior to ejaculation the produced sperm are stored in the seminal vesicle, a small gland that is located just behind the bladder.

A sperm cell is motile and swims via chemotaxis, using its flagellum to propel itself towards the ovum.

The Mammalian Female

The female reproductive system likewise contains two main divisions: the vagina and uterus, which act as the receptacle for the sperm, and the ovaries, which produce the female's ova. All of these parts are always internal. The vagina is attached to the uterus through the cervix, while the uterus is attached to the ovaries via the Fallopian tubes. At certain intervals, the ovaries release an ovum, which passes through the fallopian tube into the uterus.

If, in this transit, it meets with sperm, the sperm penetrate and merge with the egg, fertilizing it. The fertilization usually occurs in the oviducts, but can happen in the uterus itself. The zygote then implants itself in the wall of the uterus, where it begins the processes of embryogenesis and morphogenesis. When developed enough to survive outside the womb, the cervix dilates and contractions of the uterus propel the fetus through the birth canal, which is the vagina.

The ova are larger than sperm and are generally all created by birth. They are for the most part stationary, aside from their transit to the uterus, and contain nutrients for the later zygote and embryo. Over a regular interval, a process of oogenesis matures one ovum to be sent down the Fallopian tube attached to its ovary in anticipation of fertilization. If not fertilized, this egg is flushed out of the system through menstruation in humans and great apes and reabsorbed in all other mammals in the estrus cycle.

Gestation

Gestation, called *pregnancy* in humans, is the period of time during which the fetus develops, dividing via mitosis inside the female. During this time, the fetus receives all of its nutrition and oxygenated blood from the female, filtered through the placenta, which is attached to the fetus' abdomen via an umbilical cord. This drain of nutrients can be quite taxing on the female, who is required to ingest significantly higher levels of calories. In addition, certain vitamins and other nutrients are required in greater quantities than normal, often creating abnormal eating habits. The length of gestation, called the gestation period, varies greatly from species to species; it is 38 weeks in humans, 56–60 in giraffes and 16 days in hamsters.

Birth

Once the fetus is sufficiently developed, chemical signals start the process of birth, which begins with contractions of the uterus and the dilation of the cervix. The fetus then descends to the cervix, where it is pushed out into the vagina, and eventually out of the female. The newborn, which is called an infant in humans, should typically begin respiration on its own shortly after birth. Not long after, the placenta is passed as well. Most mammals eat this, as it is a good source of protein and other vital nutrients needed for caring for the young. The end of the umbilical cord attached to the young's abdomen eventually falls off on its own.

Monotremes

Monotremes, only five species of which exist, all from Australia and New Guinea, lay eggs. They have one opening for excretion and reproduction called the cloaca. They hold the eggs internally for several weeks, providing nutrients, and then lay them and cover them like birds. After less than two weeks the young hatches and crawls into its mother's pouch, much like marsupials, where it nurses for several weeks as it grows.

Marsupials

Marsupials reproduce in essentially the same manner, though their young are born at a far earlier stage of development than other mammals. After birth, marsupial joeys crawl into their mother's pouch and attach to a teat, where they receive nourishment and finish developing into self-sufficient animals.

GENETIC CODE

The genetic code is the set of rules by which information encoded in genetic material (DNA or RNA sequences) is translated into proteins (amino acid sequences) by living cells. Specifically, the code defines a mapping between tri-nucleotide sequences called codons and amino acids; every triplet of nucleotides in a nucleic acid sequence specifies a single amino acid. Because the vast majority of genes are encoded with exactly the same code, this particular code is often referred to as the canonical or standard genetic code, or simply the genetic code, though in fact there are many variant codes; thus, the canonical genetic code is not universal. For example, in humans, protein synthesis in mitochondria relies on a genetic code that varies from the canonical code.

Cracking the Genetic Code

After the structure of DNA was deciphered by James Watson, Francis Crick, Maurice Wilkins and Rosalind Franklin, serious efforts to understand the nature of the encoding of proteins began. George Gamov postulated that a three-letter code must be employed to encode the 20 different amino acids used by living cells to encode proteins (because 3 is the smallest n such that 4n is at least 20). The fact that codons did consist of three DNA bases was first demonstrated in the Crick, Brenner et al. experiment. The first elucidation of a codon was done by Marshall Nirenberg and Heinrich J. Matthaei in 1961 at the National Institutes of Health. They used a cell-free system to translate a poly-uracil RNA sequence and discovered that the polypeptide they had synthesized consisted of only the amino acid phenylalanine. They thereby deduced from this poly-phenylalanine that the codon UUU specified the amino-acid phenylalanine. Extending this work, Nirenberg and his coworkers were able to determine the nucleotide makeup of each codon. In order to determine the order of the sequence, trinucleotides were bound to ribosomes and radioactively labeled aminoacyl-tRNA was used to determine which amino acid corresponded to the codon. Nirenberg's group was able to determine the sequences of 54 out of 64 codons. Subsequent work by Har Gobind Khorana identified the rest of the code, and shortly thereafter Robert W. Holley determined the structure of transfer RNA, the adapter molecule that facilitates translation. In 1968, Khorana, Holley and Nirenberg shared the Nobel Prize in Physiology or Medicine for their work.

Transfer of Information Via the Genetic Code

The genome of an organism is inscribed in DNA, or in some viruses RNA. The portion of the genome that codes for a protein or an RNA is referred to as a gene. Those genes that code for proteins are composed of tri-nucleotide units called codons, each coding for a single amino acid. Each nucleotide sub-unit consists of a phosphate, deoxyribose sugar and one of the 4 nitrogenous nucleotide bases. The purine bases adenine (A) and guanine (G) are larger and consist of two aromatic rings. The pyrimidine bases cytosine (C) and thymine (T) are smaller and consist of only one aromatic ring. In the double-helix configuration, two strands of DNA are joined to each other by hydrogen bonds in an arrangement known as base pairing. These bonds almost always form between an adenine base on one strand and a thymine on the other strand and between a cytosine base on one strand and a guanine base on the other. This means that the number of A and T residues will be the same in a given double helix as will the number of G and C residues. In RNA, thymine (T) is replaced by uracil (U), and the deoxyribose is substituted by ribose.

Each protein-coding gene is transcribed into a template molecule of the related polymer RNA, known as messenger RNA or mRNA. This in turn is translated on the ribosome into an amino acid chain or polypeptide. The process of translation requires transfer RNAs specific for individual amino acids with the amino acids covalently attached to them, guanosine triphosphate as an energy source, and a number of translation factors. tRNAs have anticodons complementary to the codons in mRNA and can be "charged" covalently with amino acids at their 3' terminal CCA ends. Individual tRNAs are charged with specific amino acids by enzymes known as aminoacyl tRNA synthetases which have high specificity for both their cognate amino acids and tRNAs. The high specificity of these enzymes is a major reason why the fidelity of protein translation is maintained.

There are $4^3 = 64$ different codon combinations possible with a triplet codon of three nucleotides. In reality, all 64 codons of the standard genetic code are assigned for either amino acids or stop signals during translation. If, for example, an RNA sequence, UUUAAACCC is considered and the reading-frame starts with the first U (by convention, 5' to 3'), there are three codons, namely, UUU, AAA and CCC, each of which specifies one amino acid. This RNA sequence will be translated into an amino acid sequence, three amino acids long. A comparison may be made with computer science, where the codon is the equivalent of a byte, which codes for a single letter in a program (like one amino acid of a protein), and a nucleotide for a bit.

The standard genetic code is shown in the following tables, shows what amino acid each of the 64 codons specifies shows what codons specify each of the 20 standard amino acids involved in translation. These are called forward and reverse codon tables, respectively. For example, the codon AAU represents the amino acid asparagine, and UGU and UGC represent cysteine (standard three-letter designations, Asn and Cys respectively). !

Salient Features

Reading Frame of a Sequence

Note that a codon is defined by the initial nucleotide from which translation starts. For example, the string GGGAAACCC, if read from the first position, contains the codons GGG, AAA and CCC; and if read from the second position, it contains the codons GGA and AAC; if read starting from the third position, GAA and ACC. Partial codons have been ignored in this example. Every sequence can thus be read in three reading frames, each of which will produce a different amino acid sequence (in the given example, Gly-Lys-Pro, Gly-Asp, or Glu-Thr, respectively). With double-stranded DNA there are six possible reading frames, three in the forward orientation on one strand and three reverse (on the opposite strand).

The actual frame a protein sequence is translated in is defined by a start codon, usually the first AUG codon in the mRNA sequence. Mutations that disrupt the reading frame by insertions or deletions of a non-multiple of 3 nucleotide bases are known as frameshift mutations. These mutations may impair the function of the resulting protein, if it is formed, and are thus rare in *in vivo* protein-coding sequences. Often such misformed proteins are targeted for proteolytic degradation. In addition, a frame shift mutation is very likely to cause a stop codon to be read which truncates the creation of the protein. One reason for the rareness of frame-shifted mutations being inherited is that if the protein being translated is essential for growth under the selective pressures the organism faces, absence of a functional protein may cause lethality before the organism is viable.

Start/Stop Codons

Translation starts with a chain initiation codon (start codon). Unlike stop codons, the codon alone is not sufficient to begin the process. Nearby sequences and initiation factors are also required to start translation. The most common start codon is AUG, which codes for methionine, so most amino acid chains start with methionine.

The three stop codons have been given names: UAG is *amber*, UGA is *opal* (sometimes also called *umber*), and UAA is *ochre*. "Amber" was named by discoverers Richard Epstein and Charles Steinberg after their friend Harris Bernstein, whose last name means "amber" in German. The other two stop codons were named 'ochre" and "opal" in order to keep the "color names" theme. Stop codons are also called termination codons and they signal release of the nascent polypeptide from the ribosome due to binding of release factors in the absence of cognate tRNAs with anticodons complementary to these stop signals.

Degeneracy of the Genetic Code

The genetic code has redundancy but no ambiguity. For example, although codons GAA and GAG both specify glutamic acid (redundancy), neither of them specifies any other amino acid (no ambiguity). Degenerate codons may differ in their third positions; e.g., both GAA and GAG code for the amino acid glutamic acid. A codon is said to be fourfold degenerate if any nucleotide at its third position specifies the same amino acid; it is said to be twofold degenerate if only two of four possible nucleotides at its third position specify the same amino acid. In twofold degenerate codons, the equivalent third position nucleotides are always either two purines (A/G) or two pyrimidines (C/T). Only two amino acids are specified by a single codon; one of these is the amino-acid methionine, specified by the codon AUG, which also specifies the start of translation; the other is tryptophan, specified by the codon UGG. The degeneracy of the genetic code is what accounts for the existence of silent mutations.

Degeneracy results because a triplet code designates 20 amino acids and a stop codon. Because there are four bases, triplet codons are required to produce at least 21 different codes. For example, if there were two bases per codon, then only 16 amino acids could be coded for (4^2=16). Because at least 21 codes are required, then 4^3 gives 64 possible codons, meaning that some degeneracy must exist.

These properties of the genetic code make it more fault-tolerant for point mutations. For example, in theory, fourfold degenerate codons can tolerate any point mutation at the third position, although codon usage bias restricts this in practice in many organisms; twofold degenerate codons can tolerate one out of the three possible point mutations at the third position. Since transition mutations (purine to purine or pyrimidine to pyrimidine mutations) are more likely than transversion (purine to pyrimidine or vice-versa) mutations, the equivalence of purines or that of pyrimidines at twofold degenerate sites adds a further fault-tolerance.

A practical consequence of redundancy is that some errors in the genetic code only cause a silent mutation or an error that would not affect the protein because the hydrophilicity or hydrophobicity is maintained by equivalent substitution of amino acids; for example, a codon of NUN (where N = any nucleotide) tends to code for hydrophobic amino acids. Even so, it is a single point mutation that causes a modified hemoglobin molecule in sickle-cell disease. The hydrophilic glutamate (Glu) is substituted by the hydrophobic valine (Val), which reduces the solubility of â-globin. In this case, this mutation causes hemoglobin to form linear polymers linked by the hydrophobic interaction between the valine groups causing sickle-cell deformation of erythrocytes. Sickle-cell disease is generally not caused by a *de novo* mutation. Rather it is selected for in malarial regions (in a way similar to thalassemia), as heterozygous people have some resistance to the malarial *Plasmodium* parasite (heterozygote advantage).

These variable codes for amino acids are allowed because of modified bases in the first base of the anticodon of the tRNA, and the base-pair formed is called a wobble base pair. The modified bases include inosine and the Non-Watson-Crick U-G basepair.

Variations to the Standard Genetic Code

While slight variations on the standard code had been predicted earlier, none were discovered until 1979, when researchers studying human mitochondrial genes discovered they used an alternative code. Many slight variants have been discovered since, including various alternative mitochondrial codes, as well as small variants such as *Mycoplasma* translating the codon UGA as tryptophan. In bacteria and archaea, GUG and UUG are common start codons. However, in rare cases, certain specific proteins may use alternative initiation (start) codons not normally used by that species.

In certain proteins, non-standard amino acids are substituted for standard stop codons, depending upon associated signal sequences in the messenger RNA: UGA can code for selenocysteine and UAG can code for pyrrolysine as discussed in the relevant articles. Selenocysteine is now viewed as the 21st amino acid, and pyrrolysine is viewed as the 22nd. A detailed description of variations in the genetic code can be found at the NCBI web site.

However, all known codes have strong similarities to each other, and the coding mechanism is the same for all organisms: three-base codons, tRNA, and ribosomes, reading the code in the same direction, translating the code three letters at a time into sequences of amino acids.

Theories on the Origin of the Genetic Code

Despite the variations that exist, the genetic codes used by all known forms of life on Earth are very similar. Since there are many possible genetic codes that are thought to have similar utility to the one used by Earth life, the theory of evolution suggests that the genetic code was established very early in the history of life and meta-analysis of transfer RNA suggest it was established soon after the formation of earth.

One can ask the question: is the genetic code completely random, just one set of codon-amino acid correspondences that happened to establish itself and be "frozen in" early in evolution, although *functionally* any of the many other possible transcription tables would have done just as well? Already a cursory look at the table shows patterns that suggest that this is not the case.

There are three themes running through the many theories that seek to explain the evolution of the genetic code (and hence the origin of these patterns). One is illustrated by recent aptamer experiments which show that some amino acids have a selective chemical affinity for the base triplets that code for them. This suggests that the current, complex translation mechanism involving tRNA and associated enzymes may be a later development, and that originally, protein sequences were directly templated on base sequences. Another is that the standard genetic code that we see today grew from a simpler, earlier code through a process of "biosynthetic expansion". Here the idea is that primordial life 'discovered' new amino acids (e.g. as by-products of metabolism) and later back-incorporated some of these into the machinery of genetic coding. Although much circumstantial evidence has been found to suggest that fewer different amino acids were used in the past than today, precise and detailed hypotheses about exactly which amino acids entered the code in exactly what order has proved far more controversial. A third theory is that natural selection has led to codon assignments of the genetic code that minimize the effects of mutations.

REGULATION OF GENE EXPRESSION

Regulation of gene expression (or gene regulation) refers to the cellular control of the amount and timing of changes to the appearance of the functional product of a gene. Although a functional gene product may be an RNA or a protein, the majority of the known mechanisms regulate the expression of protein coding genes. Any step of the gene's expression may be modulated, from DNA-RNA transcription to the post-translational modification of a protein. Gene regulation gives the cell control over its structure and function, and is the basis for cellular differentiation, morphogenesis and the versatility and adaptability of any organism.

Regulated Stages of Gene Expression

Any step of gene expression may be modulated, from the DNA-RNA transcription step to post-translational modification of a protein. Following is a list of stages where gene expression is regulated:

- Chemical and structural modification of DNA or chromatin;
- Transcription;

- Translation;
- Post-transcriptional modification;
- RNA transport;
- mRNA degradation;
- Post-translational modifications;

Modification of DNA

Chemical

Methylation of DNA is a common method of gene silencing. DNA is typically methylated by methyltransferase enzymes on cytosine nucleotides in a CpG dinucleotide sequence (also called "CpG islands" when densely clustered). Analysis of the pattern of methylation in a given region of DNA (which can be a promoter) can be achieved through a method called bisulfite mapping. Methylated cytosine residues are unchanged by the treatment, whereas unmethylated ones are changed to uracil. The differences are analyzed by DNA sequencing or by methods developed to quantify SNPs, such as Pyrosequencing (Biotage) or MassArray (Sequenom), measuring the relative amounts of C/T at the CG dinucleotide. Abnormal methylation patterns are thought to be involved in carcinogenesis.

Structural

Transcription of DNA is dictated by its structure. In general, the density of its packing is indicative of the frequency of transcription. Octameric protein complexes called histones are responsible for the amount of supercoiling of DNA, and these complexes can be temporarily modified by processes such as phosphorylation or more permanently modified by processes such as methylation. Such modifications are considered to be responsible for more or less permanent changes in gene expression levels.

Histone acetylation is also an important process in transcription. Histone acetyltransferase enzymes (HATs) such as CREB-binding protein also dissociate the DNA from the histone complex, allowing transcription to proceed. Often, DNA methylation and histone acetylation work together in gene silencing. The combination of the two seems to be a signal for DNA to be packed more densely, lowering gene expression.

Regulation of Transcription

Regulation of transcription controls when transcription occurs and how much RNA is created. Transcription of a gene by RNA polymerase can be regulated by at least five mechanisms: (Raven 366-71)

- Specificity factors alter the specificity of RNA polymerase for a given promoter or set of promoters, making it more or less likely to bind to them (i.e. sigma factors used in prokaryotic transcription). (Raven 370)
- Repressors bind to non-coding sequences on the DNA strand that are close to or overlapping the promoter region, impeding RNA polymerase's progress along the strand, thus impeding the expression of the gene. (Raven 367)
- General transcription factors These transcription factors position RNA polymerase at the start of a protein-coding sequence and then release the polymerase to transcribe the mRNA. (Raven 370)
- Activators enhance the interaction between RNA polymerase and a particular promoter, encouraging the expression of the gene. Activators do this by increasing the attraction of RNA polymerase for the promoter, through interactions with subunits of the RNA polymerase or indirectly by changing the structure of the DNA. (Raven 369)
- Enhancers are sites on the DNA helix that are bound to by activators in order to loop the DNA bringing a specific promoter to the initiation complex.

Regulatory Protein

Regulatory protein is a term used in genetics to describe a protein involved in regulating gene expression. It is usually bound to a regulatory binding site which is sometimes located near the promotor although this is not always the case. Regulatory proteins are often needed to be bound to a regulatory binding site to switch a gene on (activator) or to shut off a gene (repressor). Generally, as the organism grows more sophisticated, their cellular protein regulation becomes more complicated and indeed some human genes can be controlled by many activators and repressors working together.

Prokaryotes vs. Eukaryotes

In prokaryotes regulation of transcription is needed for the cell to quickly adapt to the ever changing outer environment. The presence of the quantity and type of nutrients determines which genes are expressed; in order to do that, genes must be regulated in some fashion. In prokaryotes, repressors bind to regions called operators that are generally located downstream from and near the promoter (normally part of the transcript). Activators bind to the upstream portion of the promoter, such as the CAP region (completely upstream from the transcript). A combination of activators, repressors and rarely enhancers (in prokaryotes) determines whether a gene is transcribed.

In eukaryotes, transcriptional regulation tends to involve combinatorial interactions between several transcription factors, which allow for a sophisticated response to multiple conditions in the environment. This permits spatial and temporal differences in gene expression. Eukaryotes also make use of enhancers, distant regions of DNA that can loop back to the promoter. A major difference between eukaryotes and prokaryotes is the fact the eukaryotes have a nuclear envelope, which prevents simultaneous transcription and translation.

Examples:

- When E. coli bacteria are subjected to heat stress, the ó^{32} subunit of its RNA polymerase changes such that the enzyme binds to a specialized set of promoters that precede genes for heat-shock response proteins.
- When a cell contains a surplus amount of the amino acid tryptophan, the acid binds to a specialized repressor protein (tryptophan repressor). The binding changes the structural conformity of the repressor such that it binds to the operator region for the operon that synthesizes tryptophan, preventing their expression and thus suspending production. This is a form of negative feedback.
- In bacteria, the lac repressor protein blocks the synthesis of enzymes that digest lactose when there is no lactose to feed on. When lactose is present, it binds to the repressor, causing it to detach from the DNA strand.

Inducible Vs. Repressible Systems

Gene Regulation can be summarized as how they respond:

- Inducible systems - An inducible system is off unless there is the presence of some molecule (called an inducer) that allows for gene expression. The molecule is said to "induce expression". The manner in which this happens is dependent on the control mechanisms as well as differences between prokaryotic and eukaryotic cells.
- Repressible systems - A repressible system is on except in the presence of some molecule (called a corepressor) that suppresses gene expression. The molecule is said to "repress expression". The manner in which this happens is dependent on the control mechanisms as well as differences between prokaryotic and eukaryotic cells.

Regulation of Transcription Machinery

In order for a gene to be expressed, several things must happen. First, there needs to be an initiating signal. This is achieved through the binding of some ligand to a receptor. Activation of g-protein-coupled receptors can have this effect; as can the binding of hormones to intra- or extracellular receptors.

This signal gives rise to the activation of a protein called a transcription factor, and recruits other members of the "transcription machine." Transcription factors generally simultaneously bind DNA as well as an RNA polymerase, as well as other agents necessary for the transcription process (HATs, scaffolding proteins, etc.). Transcription factors, and their cofactors, can be regulated through reversible structural alterations such as phosphorylation or inactivated through such mechanisms as proteolysis.

Transcription is initiated at the promoter site, as an increase in the amount of an active transcription factor binds a target DNA sequence. Other proteins, known as "scaffolding proteins" bind other cofactors and hold them in place. DNA sequences far from the point of initiation, known as enhancers, can aid in the assembly of this "transcription machinery." Histone arms are acetylated, and DNA is transcribed into RNA.

Frequently, extracellular signals induce the expression of immediate early genes (IEGs) such as c-fos, c-jun, or AP-1. These are in and of themselves transcription factors or components thereof, and can further influence gene expression.

Posttranscriptional Regulation

After the DNA is transcribed and mRNA is formed there must be some sort of regulation on how much the mRNA is translated into Proteins. Cells do this by Capping, Splicing, and the addition of a Poly(A) Tail. These processes occur in eukaryotes but not in prokaryotes. (Raven, 374-378)

- Capping changes the five prime end of the mRNA to a three prime end by 5'-5' linkage, which protects the mRNA from 5' exonuclease, which degrades foreign RNA. The cap also helps in ribosomal binding. (Raven 378)
- Splicing removes the introns, noncoding regions that are transcribed into RNA, in order to make the mRNA able to create proteins. Cells do this by spliceosome's binding on either side of an intron, looping the intron into a circle and then cleaving it off. The two ends of the exons are then joined together. (Raven 378)
- Addition of poly(A) tail otherwise known as poly-adenylation. Junk RNA is added to the 3' end, and acts as a buffer to the 3' exonuclease in order to increase the half life of mRNA. (Raven, 378)

Up-Regulation and Down-Regulation

Up-regulation is a process which occurs within a cell triggered by a signal (originating internal or external to the cell) which results in increased expression of one or more genes and as a result the protein(s) encoded by those genes. Conversely down-regulation is a process resulting in decreased gene and corresponding protein expression.

- Up-regulation occurs for example when a cell is deficient in some kind of receptor. In this case, more receptor protein is synthesized and transported to the membrane of the cell and thus the sensitivity of the cell is brought back to normal reestablishing homeostasis.
- Down-regulation occurs for example when a cell is overly stimulated by a neurotransmitter, hormone, or drug for a prolonged period of time and the expression of the receptor protein is decreased in order to protect the cell.

MUTATION

In biology, mutations are changes to the base pair sequence of the genetic material of an organism. Mutations can be caused by copying errors in the genetic material during cell division, by exposure to ultraviolet or ionizing radiation, chemical mutagens, or viruses, or can occur deliberately under cellular control during processes such as hypermutation. In multicellular organisms, mutations can be subdivided into germ line mutations, which can be passed on to descendants, and somatic mutations, which cannot be transmitted to descendants in animals. Plants sometimes can transmit somatic mutations to their descendants asexually or sexually (in case when flower buds develop in somatically mutated part of plant). A new mutation that was not inherited from either parent is called a *de novo* mutation.

Mutations create variations in the gene pool, and the less favorable (or *deleterious*) mutations are reduced in frequency in the gene pool by natural selection, while more favorable (*beneficial* or *advantageous*) mutations tend to accumulate, resulting in evolutionary change. For example, a butterfly may produce offspring with a new mutation. Many times new mutations are harmful; a new mutation might change the color of one of the butterfly's offspring, making it harder (or easier) for predators to see. If this color change is an advantage, the chances of this butterfly surviving and producing its own offspring are a little better, and over time the number of butterflies with this mutation may form a larger percentage of the population.

Neutral mutations are defined as mutations whose effects do not influence the fitness of either the species or the individuals who make up the species. These can accumulate over time due to genetic drift. The overwhelming majority of mutations have no significant effect, since DNA repair is able to mend most changes before they become permanent mutations, and many organisms have mechanisms for eliminating otherwise permanently mutated somatic cells.

Classification

By Effect on Structure

The sequence of a gene can be altered in a number of ways. Gene mutations have varying effects on health depending on where they occur and whether they alter the function of essential proteins. Structurally, mutations can be classified as:

- Small-scale mutations, such as affecting a small gene is one or a few nucleotides, including:
  - Point mutations, often caused by chemicals or malfunction of DNA replication, exchange a single nucleotide for another. Most common is the transition that exchanges a purine for a purine (A "! G) or a pyrimidine for a pyrimidine, (C "! T). A transition can be caused by nitrous acid, base mis-pairing, or mutagenic base analogs such as 5-bromo-2-deoxyuridine (BrdU). Less common is a transversion, which exchanges a purine for a pyrimidine or a pyrimidine for a purine (C/T "! A/G). A point mutation can be reversed by another point mutation, in which the nucleotide is changed back to its original state (true reversion) or by second-site reversion (a complementary mutation elsewhere that results in regained gene functionality). These changes are classified as transitions or transversions. An example of a transversion is adenine (A) being converted into a cytosine (C). There are also many other examples that can be found. Point mutations that occur within the protein coding region of a gene may be classified into three kinds, depending upon what the erroneous codon codes for:
    * Silent mutations: which code for the same amino acid.
    * Missense mutations: which code for a different amino acid.
    * Nonsense mutations: which code for a stop and can truncate the protein.
  - Insertions add one or more extra nucleotides into the DNA. They are usually caused by transposable elements, or errors during replication of repeating elements (e.g. AT repeats). Insertions in the coding region of a gene may alter splicing of the mRNA (splice site mutation), or cause a shift in the reading frame (frameshift), both of which can significantly alter the gene product. Insertions can be reverted by excision of the transposable element.
  - Deletions remove one or more nucleotides from the DNA. Like insertions, these mutations can alter the reading frame of the gene. They are irreversible.

- Large-scale mutations in chromosomal structure, including:
  - Amplifications (or gene duplications) leading to multiple copies of all chromosomal regions, increasing the dosage of the genes located within them.
  - Deletions of large chromosomal regions, leading to loss of the genes within those regions.
  - Mutations whose effect is to juxtapose previously separate pieces of DNA, potentially bringing together separate genes to form functionally distinct fusion genes (e.g. bcr-abl). These include:
    - * Chromosomal translocations: interchange of genetic parts from nonhomologous chromosomes.
    - * Interstitial deletions: an intra-chromosomal deletion that removes a segment of DNA from a single chromosome, thereby apposing previously distant genes. For example, cells isolated from a human astrocytoma, a type of brain tumor, were found to have a chromosomal deletion removing sequences between the "fused in glioblastoma" (fig) gene and the receptor tyrosine kinase "ros", producing a fusion protein (FIG-ROS). The abnormal FIG-ROS fusion protein has constitutively active kinase activity that causes oncogenic transformation (a transformation from normal cells to cancer cells).
    - * Chromosomal inversions: reversing the orientation of a chromosomal segment.
  - Loss of heterozygosity: loss of one allele, either by a deletion or recombination event, in an organism that previously had two different alleles.

By Effect on Function

- Loss-of-function mutations are the result of gene product having less or no function. When the allele has a complete loss of function (null allele) it is often called an amorphic mutation. Phenotypes associated with such mutations are most often recessive. Exceptions are when the organism is haploid, or when the reduced dosage of a normal gene product is not enough for a normal phenotype (this is called haploinsufficiency).
- Gain-of-function mutations change the gene product such that it gains a new and abnormal function. These mutations usually have dominant phenotypes. Often called a neo-morphic mutation.
- Dominant negative mutations (also called anti-morphic mutations) have an altered gene product that acts antagonistically to the wild-type allele. These

mutations usually result in an altered molecular function (often inactive) and are characterised by a dominant or semi-dominant phenotype. In humans, Marfan syndrome is an example of a dominant negative mutation occurring in an autosomal dominant disease. In this condition, the defective glycoprotein product of the fibrillin gene (FBN1) antagonizes the product of the normal allele.

- Lethal mutations are mutations that lead to a phenotype incapable of effective reproduction.

By Aspect of Phenotype Affected

- Morphological mutations usually affect the outward appearance of an individual. Mutations can change the height of a plant or change it from smooth to rough seeds.
- Biochemical mutations result in lesions stopping the enzymatic pathway. Often, morphological mutants are the direct result of a mutation due to the enzymatic pathway.

By Inheritance

The human genome contains two copies of each gene- a paternal and a maternal allele.

- Wildtype or Homozygous non-mutated is when neither alleles are mutated.
- A Heterozygous mutation is when only one allele is mutated.
- A Homozygous mutation is when both the paternal and maternal alleles have an identical mutation.
- Compound heterozygous mutations or a Genetic compound is when the paternal and maternal alleles have two different mutations.

Special Classes

- Conditional mutation is a mutation that has wild-type (or less severe) phenotype under certain "permissive" environmental conditions and a mutant phenotype under certain "restrictive" conditions. For example, a temperature-sensitive mutation can cause cell death at high temperature (restrictive condition), but might have no deleterious consequences at a lower temperature (permissive condition).

Causes of Mutation

Two classes of mutations are spontaneous mutations (molecular decay) and induced mutations caused by mutagens.

Spontaneous mutations on the molecular level include:

- Tautomerism - A base is changed by the repositioning of a hydrogen atom.
- Depurination - Loss of a purine base (A or G).
- Deamination - Changes a normal base to an atypical base; C '! U, (which can be corrected by DNA repair mechanisms), or spontaneous deamination of 5-methycytosine (irreparable), or A '! HX (hypoxanthine).
- Transition - A purine changes to another purine, or a pyrimidine to a pyrimidine.
- Transversion - A purine becomes a pyrimidine, or vice versa.

Induced mutations on the molecular level can be caused by:

- Chemicals
  - Nitrosoguanidine (NTG)
  - Hydroxylamine NH_2OH
  - Base analogs (e.g. BrdU)
  - Simple chemicals (e.g. acids)
  - Alkylating agents (e.g. *N*-ethyl-*N*-nitrosourea (ENU)) These agents can mutate both replicating and non-replicating DNA. In contrast, a base analog can only mutate the DNA when the analog is incorporated in replicating the DNA. Each of these classes of chemical mutagens has certain effects that then lead to transitions, transversions, or deletions.
  - Methylating agents (e.g. ethyl methanesulfonate (EMS))
  - Polycyclic hydrocarbons (e.g. benzopyrenes found in internal combustion engine exhaust)
  - DNA intercalating agents (e.g. ethidium bromide)
  - DNA crosslinker (e.g. platinum)
  - Oxidative damage caused by oxygen(O)] radicals
- Radiation
  - Ultraviolet radiation (nonionizing radiation) - excites electrons to a higher energy level. DNA absorbs one form, ultraviolet light. Two nucleotide bases in DNA - cytosine and thymine-are most vulnerable to excitation that can change base-pairing properties. UV light can induce adjacent thymine bases in a DNA strand to pair with each other, as a bulky dimer.
  - Ionizing radiation

DNA has so-called hotspots, where mutations occur up to 100 times more frequently than the normal mutation rate. A hotspot can be at an unusual base, e.g., 5-methylcytosine.

Mutation rates also vary across species. Evolutionary biologists have theorized that higher mutation rates are beneficial in some situations, because they allow organisms to evolve and therefore adapt more quickly to their environments. For example, repeated exposure of bacteria to antibiotics, and selection of resistant mutants, can result in the selection of bacteria that have a much higher mutation rate than the original population (mutator strains).

Nomenclature

Nomenclature of mutations specify the type of mutation and base or amino acid changes.

- Amino acid substitution - (e.g. D111E) The first letter is the one letter code of the wildtype amino acid, the number is the position of the amino acid from the N terminus and the second letter is the one letter code of the amino acid present in the mutation. If the second letter is 'X', any amino acid may replace the wildtype.
- Amino acid deletion - (e.g. ÄF508) The Greek symbol Ä or 'delta' indicates a deletion. The letter refers to the amino acid present in the wildtype and the number is the position from the N terminus of the amino acid were it to be present as in the wildtype.

Types of Mutations

Adaptive Mutation

In mainstream biological thought it is held that while mutagenesis is non-random in many ways, the utility of a genetic mutation to the organism in which it occurs does not affect the rate at which it occurs. However experimental evidence exists that in some instances the rate of specific mutations arising is greater when they are advantageous to the organism than when they are not.

Back Mutation

Back mutation is a change in a nucleotide pair of a point-mutated DNA sequence that restores the original sequence and hence the original phenotype.

Frameshift Mutation

A *frameshift mutation* is a mutation caused by indels, ie. inserts or deletes a number of nucleotides that is not evenly divisible by three from a DNA sequence.

Due to the triplet nature of gene expression by codons, the insertion or deletion can disrupt the reading frame, or the grouping of the codons, resulting in a completely different translation from the original. The earlier in the sequence the deletion or insertion occurs, the more altered the protein produced is.

Missense Mutation

Missense mutations or *nonsynonymous mutations* are types of point mutations where a single nucleotide is changed to cause substitution of a different amino acid. This in turn can render the resulting protein nonfunctional. Such mutations are responsible for diseases such as Epidermolysis bullosa, sickle-cell disease, and SOD1 mediated ALS.

Neutral Mutation

A *neutral mutation* is a mutation that occurs in an amino acid codon (presumably within an mRNA molecule) which results in the use of a different, but chemically similar, amino acid. This is similar to a silent mutation, where a codon mutation may encode the same amino acid for example, a change from AUU to AUC will still encode leucine, so no discernable change occurs (a silent mutation).

Nonsense Mutation

A nonsense mutation is a point mutation in a sequence of DNA that results in a premature stop codon, or a *nonsense codon* in the transcribed mRNA, and possibly a truncated, and often nonfunctional protein product.

Point Mutation

A *point mutation*, or *substitution*, is a type of mutation that causes the replacement of a single base nucleotide with another nucleotide. Often the term *point mutation* also includes insertions or deletions of a single base pair (which have more of an adverse effect on the synthesized protein due to nucleotides still being read in triplets, but in different frames- a mutation called a frameshift mutation).

Silent mutation

Silent mutations are DNA mutations that do not result in a change to the amino acid sequence of a protein. They may occur in a non-coding region (outside of a gene or within an intron), or they may occur within an exon in a manner that does not alter the final amino acid sequence. The phrase silent mutation is often used interchangeably with the phrase synonymous mutation; however, synonymous mutations are a subcategory of the former, occurring only within exons.

Harmful Mutations

Changes in DNA caused by mutation can cause errors in protein sequence, creating partially or completely non-functional proteins. To function correctly, each cell depends on thousands of proteins to function in the right places at the right times. When a mutation alters a protein that plays a critical role in the body, a medical condition can result. A condition caused by mutations in one or more genes is called a genetic disorder. However, only a small percentage of mutations cause genetic disorders; most have no impact on health. For example, some mutations alter a gene's DNA base sequence but don't change the function of the protein made by the gene.

If a mutation is present in a germ cell, it can give rise to offspring that carries the mutation in all of its cells. This is the case in hereditary diseases. On the other hand, a mutation can occur in a somatic cell of an organism. Such mutations will be present in all descendants of this cell, and certain mutations can cause the cell to become malignant, and thus cause cancer.

Often, gene mutations that could cause a genetic disorder are repaired by the DNA repair system of the cell. Each cell has a number of pathways through which enzymes recognize and repair mistakes in DNA. Because DNA can be damaged or mutated in many ways, the process of DNA repair is an important way in which the body protects itself from disease.

Beneficial Mutations

A very small percentage of all mutations actually have a positive effect. These mutations lead to new versions of proteins that help an organism and its future generations better adapt to changes in their environment. For example, a specific 32 base pair deletion in human CCR5 (CCR5-Ä32) confers HIV resistance to homozygotes and delays AIDS onset in heterozygotes. The CCR5 mutation is more common in those of European descent. One theory for the etiology of the relatively high frequency of CCR5-Ä32 in the European population is that it conferred resistance to the bubonic plague in mid-14th century Europe. People who had this mutation were able to survive infection thus its frequency in the population increased. It could also explain why this mutation is not found in Africa where the bubonic plague never reached. Newer theory says the selective pressure on the CCR5 Delta 32 mutation has been caused by smallpox instead of bubonic plague.

SOMATIC CELL NUCLEAR TRANSFER

In genetics and developmental biology, somatic cell nuclear transfer (SCNT) is a laboratory technique for creating an ovum with a donor nucleus (see process below). It can be used in embryonic stem cell research, or in regenerative medicine where it is sometimes referred to as "therapeutic cloning." It can also be used as the first step in the process of reproductive cloning.

The Process

In SCNT the nucleus, which contains the organism's DNA, of a somatic cell (a body cell other than a sperm or egg cell) is removed and the rest of the cell discarded. At the same time, the nucleus of an egg cell is removed. The nucleus of the somatic cell is then inserted into the enucleated egg cell. After being inserted into the egg, the somatic cell nucleus is reprogrammed by the host cell. The egg, now containing the nucleus of a somatic cell, is stimulated with a shock and will begin to divide. After many mitotic divisions in culture, this single cell forms a blastocyst (an early stage embryo with about 100 cells) with almost identical DNA to the original organism.

SCNT in Stem Cell Research

Some researchers use SCNT in stem cell research. The aim of carrying out this procedure is to obtain stem cells that are genetically matched to the donor organism. Presently, no human stem cell lines have been derived from SCNT research.

A potential use of genetically-customized stem cells would be to create cell lines that have genes linked to the particular disease. For example, if a person with Parkinson's disease donated his or her somatic cells, then the stem cells resulting SCNT would have genes that contribute to Parkinson's disease. In this scenario, the disease-specific stem cell lines would be studied in order to better understand the disease.

In another scenario, genetically-customized stem cell lines would be generated for cell-based therapies to transplant to the patient. The resulting cells would be genetically identical to the somatic cell donor, thus avoiding any complications from immune system rejection.

Only a handful of the labs in the world are currently using SCNT techniques in human stem cell research. In the United States, scientists at the Harvard University Stem Cell Institute, the University of California San Francisco, and possibly Advanced Cell Technology are currently researching a technique to use somatic cell nuclear transfer to produce embryonic stem cells. In the United Kingdom, the Human Fertilisation and Embryology Authority has granted permission to research groups at the Roslin Institute and the Newcastle Centre for Life. SCNT may also be occurring in China.

In 2005, a South Korean research team led by Professor Hwang Woo-suk, published claims to have derived stem cell lines via SCNT, but supported those claims with fabricated data. Recent evidence has proved the he in fact created a stem cell line from a parthenote.

SCNT in Reproductive Cloning

This technique is currently the basis for cloning animals (such as the famous Dolly the sheep), and in theory could be used to clone humans. However, most researchers believe that in the foreseeable future it will not be possible to use this technique to produce a human clone that will develop to term.

Limitations

The stresses placed on both the egg cell and the introduced nucleus are enormous, leading to a high loss in resulting cells. For example, Dolly the sheep was born after 277 eggs were used for SCNT, which created 29 viable embryos. Only three of these embryos survived until birth, and only one survived to adulthood. As the procedure currently cannot be automated, but has to be performed manually under a microscope, SCNT is very resource intensive. The biochemistry involved in reprogramming the differentiated somatic cell nucleus and activating the recipient egg is also far from understood.

In SCNT, not all of the donor cell's genetic information is transferred, as the donor cell's mitochondria that contain their own mitochondrial DNA are left behind. The resulting hybrid cells retain those mitochondrial structures which originally belonged to the egg. As a consequence, clones such as Dolly that are born from SCNT are not perfect copies of the donor of the nucleus.

Controversy

Proposals to use Nucleus Transfer techniques in human stem cell research raise a set of concerns beyond the moral status of any created embryo. These have led to some individuals and organizations who are also opposed to human embryonic stem cell research to be concerned about, or opposed to, SCNT research.

One concern is that blastula creation in human stem cell research will lead to the reproductive cloning of humans. Both processes use the same first step: the creation of a nuclear transferred embryo, most likely via SCNT. Those who hold this concern often advocate for strong regulation of SCNT to preclude implantation of any derived products for the intention of human reproduction, or its prohibition.

A second concern is the appropriate source of the eggs that are needed. SCNT requires human eggs, which can only be obtained from women. The most common source of these eggs today are eggs that are produced and in excess of the clinical need during IVF treatment. This is a minimally invasive procedure, but it does carry some health risks, such as ovarian hyperstimulation syndrome, and in very rare instances even death.

Policies

SCNT is currently legal for research purposes in the United Kingdom, having been incorporated into the 1990 Human Fertilisation and Embryology Act in 2001. Permission must be obtained from the Human Fertilisation and Embryology Authority in order to perform or attempt SCNT.

In the United States, the practice remains legal, as it has not been addressed by federal law.

In 2005, the United Nations adopted a proposal submitted by Costa Rica, calling on member states to "prohibit all forms of human cloning inasmuch as they are incompatible with human dignity and the protection of human life." This phrase may include SCNT, depending on interpretation.

The Council of Europe's *Convention on Human Rights and Biomedicine* and its *Additional Protocol to the Convention for the Protection of Human Rights and Dignity of the Human Being with regard to the Application of Biology and Medicine, on the Prohibition of Cloning Human Being* appear to ban SCNT. Of the Council's 45 member states, the *Convention* has been signed by 31 and ratified by 18. The *Additional Protocol* has been signed by 29 member nations and ratified by 14.

EVOLUTION

In biology, evolution is the change in the inherited traits of a population from one generation to the next. These traits are the expression of genes that are copied and passed on to offspring during reproduction. Mutations in these genes can produce new or altered traits, resulting in heritable differences between organisms. New traits can also come from transfer of genes between populations, as in migration, or between species, in horizontal gene transfer. Evolution occurs when these heritable differences become more common or rare in a population, either non-randomly through natural selection or randomly through genetic drift.

Natural selection is a process that causes *heritable* traits that are helpful for survival and reproduction to become more common, and harmful traits to become more rare. This occurs because organisms with advantageous traits pass on more copies of these heritable traits to the next generation. Over many generations, adaptations occur through a combination of successive, small, random changes in traits, and natural selection of those variants best-suited for their environment. In contrast, genetic drift produces random changes in the frequency of traits in a population. Genetic drift arises from the role chance plays in whether a given individual will survive and reproduce.

One definition of a species is a group of organisms that can reproduce with one another and produce fertile offspring. However, when a species is separated into populations that are prevented from interbreeding, mutations, genetic drift, and the selection of novel traits cause the accumulation of differences over generations and the emergence of new species. The similarities between organisms suggest that all known species are descended from a common ancestor (or ancestral gene pool) through this process of gradual divergence.

The theory of evolution by natural selection was proposed at about the same time by both Charles Darwin and Alfred Russel Wallace, and was set out in detail in Darwin's 1859 book *On the Origin of Species*. It encountered initial resistance from religious authorities who believed humans were divinely set apart from the animal kingdom. In the 1930s, Darwinian natural selection was combined with Mendelian inheritance to form the modern evolutionary synthesis, in which the connection between the *units* of evolution (genes) and the *mechanism* of evolution (natural selection) was made. This powerful explanatory and predictive theory has become the central organizing principle of modern biology, providing a unifying explanation for the diversity of life on Earth.

Heredity

Inheritance in organisms occurs through discrete traits – particular characteristics of an organism. In humans, for example, eye color is an inherited characteristic, which individuals can inherit from one of their parents. Inherited traits are controlled by genes and the complete set of genes within an organism's genome is called its genotype.

The complete set of observable traits that make up the structure and behavior of an organism is called its phenotype. These traits come from the interaction of its genotype with the environment. As a result, not every aspect of an organism's phenotype is inherited. Suntanned skin results from the interaction between a person's genotype and sunlight; thus, a suntan is not hereditary. However, people have different responses to sunlight, arising from differences in their genotype; a striking example is individuals with the inherited trait of albinism, who do not tan and are highly sensitive to sunburn.

Genes are regions within DNA molecules that contain genetic information. DNA is a long molecule with four types of bases attached along its length. Different genes have different sequences of bases; it is the sequence of these bases that encodes genetic information. Within cells, the long strands of DNA associate with proteins to form structures called chromosomes. A specific location within a chromosome is known as a locus. If the DNA sequence at a locus varies between individuals, the different forms of this sequence are called alleles. DNA sequences can change through mutations, producing new alleles. If a mutation occurs within a gene, the new allele may affect the trait that the gene controls, altering the phenotype of the organism. However, while this simple correspondence between an allele and a trait works in some cases, most traits are more complex and are controlled by multiple interacting genes.

Variation

Because an individual's phenotype results from the interaction of their genotype with the environment, the variation in phenotypes in a population reflects the variation in these organisms' genotypes. The modern evolutionary synthesis defines evolution as the change over time in this genetic variation. The frequency of one particular allele will fluctuate, becoming more or less prevalent relative to other forms of that gene. Evolutionary forces act by driving these changes in allele frequency in one direction or another. Variation disappears when an allele reaches the point of fixation — when it either disappears from the population or replaces the ancestral allele entirely.

Variation comes from mutations in genetic material, migration between populations (gene flow), and the reshuffling of genes through sexual reproduction. Variation also comes from exchanges of genes between different species; for example, through horizontal gene transfer in bacteria, and hybridization in plants. Despite the constant introduction of variation through these processes, most of the genome of a species is identical in all individuals of that species. However, even relatively small changes in genotype can lead to dramatic changes in phenotype: chimpanzees and humans differ in only about 5% of their genomes.

Mutation

Genetic variation comes from random mutations that occur in the genomes of organisms. Mutations are changes in the DNA sequence of a cell's genome and are caused by radiation, viruses, transposons and mutagenic chemicals, as well as errors that occur during meiosis or DNA replication. These mutagens produce several different types of change in DNA sequences; these can either have no effect, alter the product of a gene, or prevent the gene from functioning. Studies in the fly *Drosophila melanogaster* suggest that about 70 percent of mutations are deleterious, and the remainder are either neutral or have a weak beneficial effect. Due to the damaging effects that mutations can have on cells, organisms have evolved mechanisms such as DNA repair to remove mutations. Therefore, the optimal mutation rate for a species is a trade-off between short-term costs, such as the risk of cancer, and the long-term benefits of advantageous mutations.

Large sections of DNA can also be duplicated, which is a major source of raw material for evolving new genes, with tens to hundreds of genes duplicated in animal genomes every million years. Most genes belong to larger families of genes of shared ancestry. Novel genes are produced either through duplication and mutation of an ancestral gene, or by recombining parts of different genes to form new combinations with new functions. For example, the human eye uses four genes to make structures that sense light: three for color vision and one for night vision; all four arose from a single ancestral gene. An advantage of duplicating a gene (or even an (entire genome) is that overlapping or redundant functions in multiple genes allows alleles to be retained that would otherwise be harmful, thus increasing genetic diversity.

Changes in chromosome number may also involve the breakage and rearrangement of DNA within chromosomes. For example, two chromosomes in the *Homo* genus fused to produce human chromosome 2; this fusion did not occur in the lineage of the other apes, and they retain these separate chromosomes. The completion of both the human genome project and the chimpanzee genome project has allowed the identification of the relevant chromosomes. In evolution, the most important role of such chromosomal rearrangements may be to accelerate the divergence of a population into new species by preserving genetic differences within populations.

Sequences of DNA that can move about the genome, such as transposons, make up a major fraction of the genetic material of plants and animals, and may have been important in the evolution of genomes. For example, more than a million copies of the Alu sequence are present in the human genome, and these sequences have now been recruited to perform functions such as regulating gene expression. Another effect of these mobile DNA sequences is that when they move within a genome, they can mutate or delete existing genes and thereby produce genetic diversity.

Recombination

In asexual organisms, genes are inherited together, or *linked*, as they cannot mix with genes in other organisms during reproduction. However, the offspring of sexual organisms contain a random mixture of their parents' chromosomes that is produced through independent assortment. In the related process of genetic recombination, sexual organisms can also exchange DNA between two matching chromosomes. These shuffling processes can allow even alleles that are close together in a strand of DNA to be inherited independently. However, as only about one recombination event occurs per million base pairs in humans, genes close together on a chromosome may not be shuffled away from each other, and tend to be inherited together. This tendency is measured by finding how often two alleles occur together, which is called their linkage disequilibrium. A set of alleles that is usually inherited in a group is called a haplotype, and this co-inheritance can indicate that the locus is under positive selection.

Recombination in sexual organisms helps to remove harmful mutations and retain beneficial mutations. Consequently, when alleles cannot be separated by recombination – such as in mammalian Y chromosomes, which pass intact from fathers to sons – harmful mutations accumulate. In addition, recombination can produce individuals with new and advantageous gene combinations. These positive effects of recombination are balanced by the fact that this process can cause mutations and separate beneficial combinations of genes. The optimal rate of recombination for a species is therefore a trade-off between conflicting factors.

Mechanisms

There are three basic mechanisms of evolutionary change: natural selection, genetic drift, and gene flow. Natural selection favors genes that improve capacity for survival and reproduction. Genetic drift is random change in the frequency of alleles, caused by the random sampling of a generation's genes during reproduction, and gene flow is the transfer of genes within and between populations. The relative importance of natural selection and genetic drift in a population varies depending on the strength of the selection and the effective population size, which is the number of individuals capable of breeding. Natural selection usually predominates in large populations, while genetic drift dominates in small populations. The dominance of genetic drift in small populations can even lead to the fixation of slightly deleterious mutations. As a result, changing population size can dramatically influence the course of evolution. Population bottlenecks, where the population shrinks temporarily and therefore loses genetic variation, result in a more uniform population. Bottlenecks also result from alterations in gene flow such as decreased migration, expansions into new habitats, or population subdivision.

Natural Selection

Natural selection is the process by which genetic mutations that enhance reproduction become, and remain, more common in successive generations of a population. It has often been called a "self-evident" mechanism because it necessarily follows from three simple facts:

- Heritable variation exists within populations of organisms.
- Organisms produce more offspring than can survive.
- These offspring vary in their ability to survive and reproduce.

These conditions produce competition between organisms for survival and reproduction. Consequently, organisms with traits that give them an advantage over their competitors pass these advantageous traits on, while traits that do not confer an advantage are not passed on to the next generation.

The central concept of natural selection is the evolutionary fitness of an organism. This measures the organism's genetic contribution to the next generation. However, this is not the same as the total number of offspring: instead fitness measures the proportion of subsequent generations that carry an organism's genes. Consequently, if an allele increases fitness more than the other alleles of that gene, then with each generation this allele will become more common within the population. These traits are said to be "selected *for*". Examples of traits that can increase fitness are enhanced survival, and increased fecundity. Conversely, the lower fitness caused by having a less beneficial or deleterious allele results in this allele becoming rarer — they are "selected *against*". Importantly, the fitness of an allele is not a fixed characteristic, if the environment changes, previously neutral or harmful traits may become beneficial and previously beneficial traits become harmful.

Natural selection within a population for a trait that can vary across a range of values, such as height, can be categorized into three different types. The first is directional selection, which is a shift in the average value of a trait over time — for example organisms slowly getting taller. Secondly, disruptive selection is selection for extreme trait values and often results in two different values becoming most common, with selection against the average value. This would be when either short or tall organisms had an advantage, but not those of medium height. Finally, in stabilizing selection there is selection against extreme trait values on both ends, which causes a decrease in variance around the average value. This would, for example, cause organisms to slowly become all the same height.

A special case of natural selection is sexual selection, which is selection for any trait that increases mating success by increasing the attractiveness of an organism to potential mates. Traits that evolved through sexual selection are particularly prominent in males of some animal species, despite traits such as cumbersome antlers, mating calls or bright colors that attract predators, decreasing the survival of individual males. This survival disadvantage is balanced by higher reproductive success in males that show these hard to fake, sexually selected traits.

An active area of research is the unit of selection, with natural selection being proposed to work at the level of genes, cells, individual organisms, groups of organisms and even species. None of these models are mutually-exclusive and selection may act on multiple levels simultaneously. Below the level of the individual, genes called transposons try to copy themselves throughout the genome. Selection at a level above the individual, such as group selection, may allow the evolution of co-operation, as discussed below.

Genetic Drift

Genetic drift is the change in allele frequency from one generation to the next that occurs because alleles in the offspring generation are a random sample of those in the parent generation, and are thus subject to sampling error. As a result, when selective forces are absent or relatively weak, allele frequencies tend to "drift" upward or downward in a random walk. This drift halts when an allele eventually becomes fixed, either by disappearing from the population, or replacing the other alleles entirely. Genetic drift may therefore eliminate some alleles from a population due to chance alone, and two separate populations that began with the same genetic structure can drift apart by random fluctuation into two divergent populations with different sets of alleles. The time for an allele to become fixed by genetic drift depends on population size, with fixation occurring more rapidly in smaller populations.

Although natural selection is responsible for adaptation, the relative importance of the two forces of natural selection and genetic drift in driving evolutionary change in general is an area of current research in evolutionary biology. These investigations were prompted by the neutral theory of molecular evolution, which proposed that most evolutionary changes are the result the fixation of neutral mutations that do not have any immediate effects on the fitness of an organism. Hence, in this model, most genetic changes in a population are the result of constant mutation pressure and genetic drift.

Gene Flow

Gene flow is the exchange of genes between populations, which are usually of the same species. Examples of gene flow within a species include the migration and then breeding of organisms, or the exchange of pollen. Gene transfer between species includes the formation of hybrid organisms and horizontal gene transfer.

Migration into or out of a population can change allele frequencies. Immigration may add new genetic material to the established gene pool of a population. Conversely, emigration may remove genetic material. As barriers to reproduction between two diverging populations are required for the populations to become new species, gene flow may slow this process by spreading genetic differences between the populations. Gene flow is hindered by mountain ranges, oceans and deserts or even man-made structures such as the Great Wall of China, which has hindered the flow of plant genes.

Depending on how far two species have diverged since their most recent common ancestor, it may still be possible for them to produce offspring, as with horses and donkeys mating to produce mules. Such hybrids are generally infertile, due to the two different sets of chromosomes being unable to pair up during meiosis. In this case, closely-related species may regularly interbreed, but hybrids will be selected against and the species will remain distinct. However, viable hybrids are occasionally formed and these new species can either have properties intermediate between their parent species, or possess a totally new phenotype. The importance of hybridization in creating new species of animals is unclear, although cases have been seen in many types of animals, with the gray tree frog particularly well-studied.

Hybridization is, however, an important means of speciation in plants, since polyploidy (having more than two copies of each chromosome) is tolerated in plants more readily than in animals. Polyploidy is important in hybrids as it allows reproduction, with the two different sets of chromosomes each being able to pair with an identical partner during meiosis. Polyploids also have more genetic diversity, which allows them to avoid inbreeding depression in small populations.

Horizontal gene transfer is the transfer of genetic material from one organism to another organism that is not its offspring, this is most common among bacteria. In medicine, this contributes to the spread of antibiotic resistance, as when one bacteria acquires resistance genes it can rapidly transfer them to other species. Horizontal transfer of genes from bacteria to eukaryotes such as the yeast *Saccharomyces cerevisiae* and the adzuki bean beetle *Callosobruchus chinensis* may also have occurred. Viruses can also carry DNA between organisms, allowing transfer of genes even across biological domains. Gene transfer has also occurred within eukaryotic cells, from the chloroplast and mitochondrial genomes to nuclear genomes.

Outcomes

Evolution influences every aspect of the form and behavior of organisms. Most prominent are the specific behavioral and physical adaptations that are the outcome of natural selection. These adaptations increase fitness by aiding activities such as finding food, avoiding predators or attracting mates. Organisms can also respond to selection by co-operating with each other, usually by aiding their relatives or engaging in mutually-beneficial symbiosis. In the longer term, evolution produces new species through splitting ancestral populations of organisms into new groups that are unable to breed with one another.

These outcomes of evolution are sometimes divided into macroevolution, which is evolution that occurs at or above the level of species, such as speciation, and microevolution, which is smaller evolutionary changes, such as adaptations, within a species or population. In general, macroevolution is the outcome of long periods of microevolution. Thus, the distinction between micro- and macroevolution is not a fundamental one - the difference is simply the time involved. However, in macroevolution, the traits of the entire species are important. For instance, a large amount of variation among individuals allows a species to rapidly adapt to new habitats, lessening the chance of it going extinct, while a wide geographic range increases the chance of speciation, by making it more likely that part of the population will become isolated. In this sense, microevolution and macroevolution can sometimes be separate.

A common misconception is that evolution is "progressive," but natural selection has no long-term goal and does not necessarily produce greater complexity. Although complex species have evolved, this occurs as a side effect of the overall number of organisms increasing, and simple forms of life remain more common. For example, the overwhelming majority of species are microscopic prokaryotes, which form about half the world's biomass despite their small size, and constitute the vast majority of Earth's biodiversity. Simple organisms therefore remain the dominant form of life on Earth, and complex life appears more diverse only because it is more noticeable.

Adaptation

Adaptations are structures or behaviors that enhance a specific function, causing organisms to become better at surviving and reproducing. They are produced by a combination of the continuous production of small, random changes in traits, followed by natural selection of the variants best-suited for their environment. This process can cause either the gain of a new feature, or the loss of an ancestral feature. An example that shows both types of change is bacterial adaptation to antibiotic selection, with mutations causing antibiotic resistance by either modifying the target of the drug, or removing the transporters that allow the drug into the cell. However, many traits that appear to be simple adaptations are in fact exaptations: structures originally adapted for one function, but which coincidentally became somewhat useful for some other function in the process. One example is the African lizard *Holapsis guentheri*, which developed an extremely flat head for hiding in crevices, as can be seen by looking at its near relatives. However, in this species, the head has become so flattened that it assists in gliding from tree to tree - an exaptation.

As adaptation occurs through the gradual modification of existing structures, structures with similar internal organization may have very different functions in related organisms. This is the result of a single ancestral structure being adapted to function in different ways. The bones within bat wings, for example, are structurally similar to both human hands and seal flippers, due to the common descent of these structures from an ancestor that also had five digits at the end of each forelimb. Other idiosyncratic anatomical features, such as bones in the wrist of the panda being formed into a false "thumb," indicate that an organism's evolutionary lineage can limit what adaptations are possible. During adaption, some structures may lose their original function and become vestigial structures. Such structures may have little or no function in a current species, yet have a clear function in ancestral species, or other closely-related species. Examples include the non-functional remains of eyes in blind cave-dwelling fish, wings in flightless birds, and the presence of hip bones in whales and snakes. Examples of vestigial structures in humans include wisdom teeth, the coccyx, and the vermiform appendix.

An area of current investigation in evolutionary developmental biology is the developmental basis of adaptations and exaptations. This research addresses the origin and evolution of embryonic development and how modifications of development and developmental processes produce novel features. These studies have shown that evolution can alter development to create new structures, such as embryonic bone structures that develop into the jaw in other animals instead forming part of the middle ear in mammals. It is also possible for structures that have been lost in evolution to reappear due to changes in developmental genes, such as a mutation in chickens causing embryos to grow teeth similar to those of crocodiles.

Co-Evolution

Interactions between organisms can produce both conflict and co-operation. When the interaction is between pairs of species, such as a pathogen and a host, or a predator and its prey, these species can develop matched sets of adaptations. Here, the evolution of one species causes adaptations in a second species. These changes in the second species then, in turn, cause new adaptations in the first species. This cycle of selection and response is called co-evolution. An example is the production of tetrodotoxin in the rough-skinned newt and the evolution of tetrodotoxin resistance in its predator, the common garter snake. In this predator-prey pair, an evolutionary arms race has produced high levels of toxin in the newt and correspondingly high levels of resistance in the snake.

Co-Operation

However, not all interactions between species involve conflict. Many cases of mutually beneficial interactions have evolved. For instance, an extreme cooperation exists between plants and the mycorrhizal fungi that grow on their roots and aid the plant in absorbing nutrients from the soil. This is a reciprocal relationship as the plants provide the fungi with sugars from photosynthesis. Here, the fungi actually grow inside plant cells, allowing them to exchange nutrients with their hosts, while sending signals that suppress the plant immune system.

Coalitions between organisms of the same species have also evolved. An extreme case is the Eusociality found in social insects, such as bees, termites and ants, where sterile insects feed and guard the small number of organisms in a colony that are able to reproduce. On an even smaller scale, the somatic cells that make up the body of an animal are limited in their capacity to reproduce in order to maintain a stable organism which then supports a small number of the animal's germ cells to produce offspring. Here, somatic cells respond to specific signals that instruct them to either grow or kill themselves. If cells ignore these signals and attempt to multiply inappropriately, their uncontrolled growth causes cancer.

These examples of cooperation within species are thought to have evolved through the process of kin selection, which is where one organism acts to help raise a relative's offspring. This activity is selected for because if the *helping* individual contains alleles which promote the helping activity, it is likely that its kin will *also* contain these alleles and thus those alleles will be passed on. Other processes that may promote cooperation include group selection, where cooperation provides benefits to a group of organisms.

Speciation

Speciation is the process where a species diverges into two or more descendant species. It has been observed multiple times under both controlled laboratory conditions and in nature. In sexually-reproducing organisms, speciation results from reproductive isolation followed by genealogical divergence. There are four mechanisms for speciation. The most common in animals is allopatric speciation, which occurs in populations initially isolated geographically, such as by habitat fragmentation or migration. As selection and drift act independently in isolated populations, separation will eventually produce organisms that cannot interbreed.

The second mechanism of speciation is peripatric speciation, which occurs when small populations of organisms become isolated in a new environment. This differs from allopatric speciation in that the isolated populations are numerically much smaller than the parental population. Here, the founder effect causes rapid speciation through both rapid genetic drift and selection on a small gene pool.

The third mechanism of speciation is parapatric speciation. This is similar to peripatric speciation in that a small population enters a new habitat, but differs in that there is no physical separation between these two populations. Instead, speciation results from the evolution of mechanisms that reduce gene flow between the two populations. Generally this occurs when there has been a drastic change in the environment within the parental species' habitat. One example is the grass *Anthoxanthum odoratum*, which can undergo parapatric speciation in response to localized metal pollution from mines. Here, plants evolve that have resistance to high levels of metals in the soil. Selection against interbreeding with the metal-sensitive parental population produces a change in flowering time of the metal-resistant plants, causing reproductive isolation. Selection against hybrids between the two populations may cause *reinforcement*, which is the evolution of traits that promote mating within a species, as well as character displacement, which is when two species become more distinct in appearance.

Finally, in sympatric speciation species diverge without geographic isolation or changes in habitat. This form is rare since even a small amount of gene flow may remove genetic differences between parts of a population. Generally, sympatric speciation in animals requires the evolution of both genetic differences and non-random mating, to allow reproductive isolation to evolve.

One type of sympatric speciation involves cross-breeding of two related species to produce a new hybrid species. This is not common in animals as animal hybrids are usually sterile, because during meiosis the homologous chromosomes from each parent, being from different species cannot successfully pair. It is more common in plants, however because plants often double their number of chromosomes, to form polyploids. This allows the chromosomes from each parental species to form a matching pair during meiosis, as each parent's chromosomes is represented by a pair already. An example of such a speciation event is when the plant *Oenothera lamarckiana* gave rise to a new species of plant *O. gigas* in 1905, which had a different chromosome number than its parents. (O. gigas had a chromosome number of 2N = 28 while its parent *Oenothera lamarckiana* had a chromosome number of 2N = 14.)

Indeed, chromosome doubling can itself cause reproductive isolation, as half the doubled chromosomes will be unmatched when breeding with undoubled organisms.

Speciation events are important in the theory of punctuated equilibrium, which accounts for the pattern in the fossil record of short "bursts" of evolution interspersed with relatively long periods of stasis, where species remain relatively unchanged. In this theory, speciation and rapid evolution are linked, with natural selection and genetic drift acting most strongly on organisms undergoing speciation in novel habitats or small populations. As a result, the periods of stasis in the fossil record correspond to the parental population, and the organisms undergoing speciation and rapid evolution are found in small populations or geographically-restricted habitats, and therefore rarely being preserved as fossils.

Extinction

Extinction is the disappearance of an entire species. Extinction is not an unusual event, as species regularly appear through speciation, and disappear through extinction. Indeed, virtually all animal and plant species that have lived on earth are now extinct. These extinctions have happened continuously throughout the history of life, although the rate of extinction spikes in occasional mass extinction events. The Cretaceous–Tertiary extinction event, during which the dinosaurs went extinct, is the most well-known, but the earlier Permian-Triassic extinction event was even more severe, with approximately 96 percent of species driven to extinction. The Holocene extinction event is an ongoing mass extinction associated with humanity's expansion across the globe over the past few thousand years. Present-day extinction rates are 100-1000 times greater than the background rate, and up to 30 percent of species may be extinct by the mid 21st century. Human activities are now the primary cause of the ongoing extinction event; global warming may further accelerate it.

The role of extinction in evolution depends on which type is considered. The causes of the continuous "low-level" extinction events, which form the majority of extinctions, are not well understood and may be the result of competition between species for shared resources. If competition from other species does alter the probability that a species will become extinct, this could produce species selection as a level of natural selection. The intermittent mass extinctions are also important, but instead of acting as a selective force, they drastically reduce diversity in a nonspecific manner and promote bursts of rapid evolution and speciation in survivors.

Evolutionary History of Life

Origin of Life

The origin of life is a necessary precursor for biological evolution, but understanding that evolution occurred once organisms appeared and investigating how this happens, does not depend on understanding exactly how life began. The current scientific consensus is that the complex biochemistry that makes up life came from simpler chemical reactions, but it is unclear how this occurred. Not much is certain about the earliest developments in life, the structure of the first living things, or the identity and nature of any last universal common ancestor or ancestral gene pool. Consequently, there is no scientific consensus on how life began, but proposals include self-replicating molecules such as RNA, and the assembly of simple cells.

Common Descent

All organisms on Earth are descended from a common ancestor or ancestral gene pool. Current species are a stage in the process of evolution, with their diversity the product of a long series of speciation and extinction events. The common descent of organisms was first deduced from four simple facts about organisms: First, they have geographic distributions that cannot be explained by local adaptation. Second, the diversity of life is not a set of completely unique organisms, but organisms that share morphological similarities. Third, vestigial traits with no clear purpose resemble functional ancestral traits, and finally, that organisms can be classified using these similarities into a hierarchy of nested groups.

Past species have also left records of their evolutionary history. Fossils, along with the comparative anatomy of present-day organisms, constitute the morphological, or anatomical, record. By comparing the anatomies of both modern and extinct species, paleontologists can infer the lineages of those species. However, this approach is most successful for organisms that had hard body parts, such as shells, bones or teeth. Further, as prokaryotes such as bacteria and archaea share a limited set of common morphologies, their fossils do not provide information on their ancestry.

More recently, evidence for common descent has come from the study of biochemical similarities between organisms. For example, all living cells use the same nucleic acids and amino acids. The development of molecular genetics has revealed the record of evolution left in organisms' genomes: dating when species diverged through the molecular clock produced by mutations. For example, these DNA sequence comparisons have revealed the close genetic similarity between humans and chimpanzees and shed light on when the common ancestor of these species existed.

Evolution of Life

Despite the uncertainty on how life began, it is clear that prokaryotes were the first organisms to inhabit Earth, approximately 3–4 billion years ago. No obvious changes in morphology or cellular organization occurred in these organisms over the next few billion years.

The eukaryotes were the next major innovation in evolution. These came from ancient bacteria being engulfed by the ancestors of eukaryotic cells, in a cooperative association called endosymbiosis. The engulfed bacteria and the host cell then underwent co-evolution, with the bacteria evolving into either mitochondria or hydrogenosomes. An independent second engulfment of cyanobacterial-like organisms led to the formation of chloroplasts in algae and plants.

The history of life was that of the unicellular eukaryotes, prokaryotes, and archaea until about a billion years ago when multicellular organisms began to appear in the oceans in the Ediacaran period. The evolution of multicellularity occurred in multiple independent events, in organisms as diverse as sponges, brown algae, cyanobacteria, slime moulds and myxobacteria.

Soon after the emergence of these first multicellular organisms, a remarkable amount of biological diversity appeared over approximately 10 million years, in an event called the Cambrian explosion. Here, the majority of types of modern animals evolved, as well as unique lineages that subsequently became extinct. Various triggers for the Cambrian explosion have been proposed, including the accumulation of oxygen in the atmosphere from photosynthesis. About 500 million years ago, plants and fungi colonized the land, and were soon followed by arthropods and other animals. Amphibians first appeared around 300 million years ago, followed by early amniotes, then mammals around 200 million years ago and birds around 100 million years ago (both from "reptile"-like lineages). However, despite the evolution of these large animals, smaller organisms similar to the types that evolved early in this process continue to be highly successful and dominate the Earth, with the majority of both biomass and species being prokaryotes.

History of Evolutionary Thought

Evolutionary ideas such as common descent and the transmutation of species have existed since at least the 6th century BC, when they were expounded by the Greek philosopher Anaximander. Evolutionary thought was further developed by other early thinkers, including the Greek philosopher Empedocles, the Roman philosopher Lucretius, the Arab biologist Al-Jahiz, the Persian philosopher Ibn Miskawayh, and the Brethren of Purity. Also in the Far East, the philosopher Zhuangzi discussed a transformative power of species to adapt to their surroundings. As biological knowledge grew in the 18th century, a variety of such ideas developed, beginning with Pierre Maupertuis in 1745, and with contributions from natural philosophers such as Erasmus Darwin and Jean-Baptiste Lamarck. In 1858, Charles Darwin and Alfred Russel Wallace jointly proposed the theory of evolution by natural selection to the Linnean Society of London in separate papers. Shortly after, Darwin's publication of *On the Origin of Species* provided detailed support for the theory and led to increasingly wide acceptance of the occurrence of evolution.

Nonetheless, Darwin's specific ideas about evolution, such as gradualism and the mechanisms of natural selection, were strongly contested at first. Lamarckists argued that transmutation of species occurred as parents passed on adaptations acquired during their lifetimes. Eventually, when experiments failed to support it, this idea was abandoned in favor of Darwinism. More significantly, Darwin could not account for how traits were passed down from generation to generation. A mechanism was provided in 1365 by Gregor Mendel, who found that traits were inherited in a predictable manner. When Mendel's work was rediscovered in 1900, disagreements over the rate of evolution predicted by early geneticists and biometricians led to a rift between the Mendelian and Darwinian models of evolution.

This contradiction was reconciled in the 1930s by biologists such as Ronald Fisher. The end result was a combination of evolution by natural selection and Mendelian inheritance, the modern evolutionary synthesis. In the 1940s, the identification of DNA as the genetic material by Oswald Avery and colleagues and the subsequent publication of the structure of DNA by James Watson and Francis Crick in 1953, demonstrated the physical basis for inheritance. Since then, genetics and molecular biology have become core parts of evolutionary biology and have revolutionized the field of phylogenetics.

In its early history, evolutionary biology primarily drew in scientists from traditional taxonomically-oriented disciplines, whose specialist training in particular organisms addressed general questions in evolution. As evolutionary biology expanded as an academic discipline, particularly after the development of the modern evolutionary synthesis, it began to draw more widely from the biological sciences. Currently the study of evolutionary biology involves scientists from fields as diverse as biochemistry, ecology, genetics and physiology, and evolutionary concepts are used in even more distant disciplines such as psychology, medicine, philosophy and computer science.

Social and Cultural Views

Even before the publication of *On the Origin of Species*, the idea that life had evolved was a source of debate and evolution is still a contentious concept. Debate has generally centered on the philosophical, social and religious implications of evolution, not on the science itself; the proposition that biological evolution occurs through the mechanism of natural selection is standard in the scientific literature.

As Darwin recognized early on, the most controversial aspect of evolutionary thought is its implications to humans. Specifically, some people believe that the origin of humans involved supernatural intervention, rather than natural processes. Although many religions and denominations have reconciled their beliefs with evolution through theistic evolution, several denominations contain creationists who object to evolution, as it contradicts their literal interpretation of origin beliefs. In some countries – notably the United States – these tensions between scientific and religious teachings have fueled the ongoing creation–evolution controversy, a religious conflict focusing on politics and public education. While other scientific fields such as cosmology and earth science also conflict with literal interpretations of many religious texts, evolutionary biology has strong oppositions from believers of religion.

Evolution has been used to support philosophical positions that promote discrimination and racism. For example, the eugenic ideas of Francis Galton were developed to argue that the human gene pool should be improved by selective breeding policies, including incentives for those considered "good stock" to reproduce, and the compulsory sterilization, prenatal testing, birth control, and even killing, of those considered *bad stock*. Another example of an extension of evolutionary theory that is now widely regarded as unwarranted is "Social Darwinism," a term given to the 19th century Whig Malthusian theory developed by Herbert Spencer into ideas about "survival of the fittest" in commerce and human societies as a whole, and by others into claims that social inequality, racism, and imperialism were justified. However, contemporary scientists and philosophers consider these ideas to have been neither mandated by evolutionary theory nor supported by data.

Applications in Technology

A major technological application of evolution is artificial selection, which is the intentional selection of certain traits in a population of organisms. Humans have used artificial selection for thousands of years in the domestication of plants and animals. More recently, such selection has become a vital part of genetic engineering, with selectable markers such as antibiotic resistance genes being used to manipulate DNA in molecular biology.

As evolution can produce highly optimized processes and networks, it has many applications in computer science. Here, simulations of evolution using evolutionary algorithms and artificial life started with the work of Nils Aall Barricelli in the 1960s, and was extended by Alex Fraser, who published a series of papers on simulation of artificial selection. Artificial evolution became a widely recognized optimization method as a result of the work of Ingo Rechenberg in the 1960s and early 1970s, who used evolution strategies to solve complex engineering problems. Genetic algorithms in particular became popular through the writing of John Holland. As academic interest grew, dramatic increases in the power of computers allowed practical applications. Evolutionary algorithms are now used to solve multi-dimensional problems more efficiently than software produced by human designers, and also to optimize the design of systems.

REFERENCES

Adleman L (1994). "Molecular computation of solutions to combinatorial problems". Science 266 (5187): 1021–4.

Agris P (2004). "Decoding the genome: a modified view". Nucleic Acids Res 32 (1): 223–38.

Alberts, Bruce; Alexander Johnson, Julian Lewis, Martin Raff, Keith Roberts, and Peter Walters (2002). Molecular Biology of the Cell; Fourth Edition. New York and London: Garland Science.

Allemand, et al (1998). "Stretched and overwound DNA forms a Pauling-like structure with exposed bases". PNAS 24: 14152-14157.

Amirnovin R. (1997) An analysis of the metabolic theory of the origin of the genetic code. Journal of Molecular Evolution 44(5), 473-6.

Ashish Gehani, Thomas LaBean and John Reif. DNA-Based Cryptography. Proceedings of the 5th DIMACS Workshop on DNA Based Computers, Cambridge, MA, USA, 14–15 June 1999.

Astbury W, (1947). "Nucleic acid". Symp. SOC. Exp. Bbl 1 (66).

Baldi, Pierre. Brunak, Soren. Bioinformatics: The Machine Learning Approach MIT Press (2001)

Basu H, Feuerstein B, Zarling D, Shafer R, Marton L (1988). "Recognition of Z-RNA and Z-DNA determinants by polyamines in solution: experimental and theoretical studies". J Biomol Struct Dyn 6 (2): 299–309.

Benham C, Mielke S (2005). "DNA mechanics". Annu Rev Biomed Eng 7: 21–53.

Berg J., Tymoczko J. and Stryer L. (2002) Biochemistry. W. H. Freeman and Company.

Berns K (1990). "Parvovirus replication". Microbiol Rev 54 (3): 316–29.

Bhattacharya, Shaoni. "Killer convicted thanks to relative's DNA". newscientist.com (20 April 2004). Accessed 22 Dec 06

Bickle T, Krüger D (1993). "Biology of DNA restriction". Microbiol Rev 57 (2): 434–50.

Bird A (2002). "DNA methylation patterns and epigenetic memory". Genes Dev 16 (1): 6–21.

Bogin, Barry, Edward Alcamo, Curtis Chubb, William J. Ehmann, Mark R. Feil, David R. Hershey, Mitchell Leslie, Karel F. Liem, William Thwaites, and Salvatore Tocci. Austin: Holt, Rinehart, and Winston, 1999. 146.

Bogin, Barry, Edward Alcamo, Curtis Chubb, William J. Ehmann, Mark R. Feil, David R. Hershey, Mitchell Leslie, Karel F. Liem, William Thwaites, and Salvatore Tocci. Austin: Holt, Rinehart, and Winston, 1999. 146.

Braña M, Cacho M, Gradillas A, de Pascual-Teresa B, Ramos A (2001). "Intercalators as anticancer drugs". Curr Pharm Des 7 (17): 1745–80.

Burge S, Parkinson G, Hazel P, Todd A, Neidle S (2006). "Quadruplex DNA: sequence, topology and structure". Nucleic Acids Res 34 (19): 5402–15.

Butler, John M. (2001) Forensic DNA Typing "Elsevier". pp. 14–15.

Cadet J, Delatour T, Douki T, Gasparutto D, Pouget J, Ravanat J, Sauvaigo S (1999). "Hydroxyl radicals and DNA base damage". Mutat Res 424 (1–2): 9–21.

Cathcart R, Schwiers E, Saul R, Ames B (1984). "Thymine glycol and thymidine glycol in human and rat urine: a possible assay for oxidative DNA damage". Proc Natl Acad Sci U S A 81 (18): 5633–7.

Chalikian T, Völker J, Plum G, Breslauer K (1999). "A more unified picture for the thermodynamics of nucleic acid duplex melting: a characterization by calorimetric and volumetric techniques". Proc Natl Acad Sci U S A 96 (14): 7853–8.

Champoux J (2001). "DNA topoisomerases: structure, function, and mechanism". Annu Rev Biochem 70: 369–413.

Choudhuri S (2004). "Gene Regulation and Molecular Toxicology". Toxicology Mechanisms and Methods 15 (1): 1-23.

Ciudad J, Cid E, Velasco A, Lara JM, Aijón J, Orfao A (2002). "Flow cytometry measurement of the DNA contents of G0/G1 diploid cells from three different teleost fish species". Cytometry 48 (1): 20–5.

Clausen-Schaumann H, Rief M, Tolksdorf C, Gaub H (2000). "Mechanical stability of single DNA molecules". Biophys J 78 (4): 1997–2007.

Colin Pitchfork — first murder conviction on DNA evidence also clears the prime suspect Forensic Science Service Accessed 23 Dec 2006

Crick, F.H.C. On degenerate templates and the adaptor hypothesis (PDF). genome.wellcome.ac.uk (Lecture, 1955). Accessed 22 Dec 2006

Dahm R (2005). "Friedrich Miescher and the discovery of DNA". Dev Biol 278 (2): 274–88.

Dame RT (2005). "The role of nucleoid-associated proteins in the organization and compaction of bacterial chromatin". Mol. Microbiol. 56 (4): 858-70.

Davenport R (2001). "Ribozymes. Making copies in the RNA world". Science 292 (5520): 1278.

Davies J, Stanley J (1989). "Geminivirus genes and vectors". Trends Genet 5 (3): 77–81.

Dickman M, Ingleston S, Sedelnikova S, Rafferty J, Lloyd R, Grasby J, Hornby D (2002). "The RuvABC resolvasome". Eur J Biochem 269 (22): 5492–501.

DNA Identification in Mass Fatality Incidents. National Institute of Justice (September 2006).

Doherty A, Suh S (2000). "Structural and mechanistic conservation in DNA ligases.". Nucleic Acids Res 28 (21): 4051–8.

Douki T, Reynaud-Angelin A, Cadet J, Sage E (2003). "Bipyrimidine photoproducts rather than oxidative lesions are the main type of DNA damage involved in the genotoxic effect of solar UVA radiation". Biochemistry 42 (30): 9221–6.

Dubcovsky J., Luo MC, Zhong GY, et al (1996). "Genetic map of diploid wheat, Triticum monococcum L., and its comparison with maps of Hordeum vulgare L". Genetics 143 (2): 983–99.

Ellis NA, Ciocci S, German J (2001). "Back mutation can produce phenotype reversion in Bloom syndrome somatic cells". Hum Genet 108 (2): 167-73.

Ferguson L, Denny W (1991). "The genetic toxicology of acridines". Mutat Res 258 (2): 123–60.

Freeland S.J.; Wu T. and Keulmann N. (2003) The Case for an Error Minimizing Genetic Code. Orig Life Evol Biosph. 33(4-5), 457-77.

Galvani A, Slatkin M (2003). "Evaluating plague and smallpox as historical selective pressures for the CCR5-Ä32 HIV-resistance allele". Proc Natl Acad Sci U S A 100 (25): 15276-9.

Genetic Code page in the NCBI Taxonomy section (Downloaded 27 April 2007.)

Ghosh A, Bansal M (2003). "A glossary of DNA structures from A to Z". Acta Crystallogr D Biol Crystallogr 59 (Pt 4): 620–6.

Gommers-Ampt J, Van Leeuwen F, de Beer A, Vliegenthart J, Dizdaroglu M, Kowalak J, Crain P, Borst P (1993). "beta-D-glucosyl-hydroxymethyluracil: a novel modified base present in the DNA of the parasitic protozoan T. brucei". Cell 75 (6): 1129–36.

Gregory S, et al. (2006). "The DNA sequence and biological annotation of human chromosome 1". Nature 441 (7091): 315–21.

Gregory T (2005). "The C-value enigma in plants and animals: a review of parallels and an appeal for partnership". Ann Bot (Lond) 95 (1): 133–46.

Greider C, Blackburn E (1985). "Identification of a specific telomere terminal transferase activity in Tetrahymena extracts". Cell 43 (2 Pt 1): 405–13.

Griffith J, Comeau L, Rosenfield S, Stansel R, Bianchi A, Moss H, de Lange T (1999). "Mammalian telomeres end in a large duplex loop". Cell 97 (4): 503–14.

Grosschedl R, Giese K, Pagel J (1994). "HMG domain proteins: architectural elements in the assembly of nucleoprotein structures". Trends Genet 10 (3): 94–100.

Gusfield, Dan. Algorithms on Strings, Trees, and Sequences: Computer Science and Computational Biology. Cambridge University Press, 15 January 1997.

Harrison P, Gerstein M (2002). "Studying genomes through the aeons: protein families, pseudogenes and proteome evolution". J Mol Biol 318 (5): 1155–74.

Harrison P, Hegyi H, Balasubramanian S, Luscombe N, Bertone P, Echols N, Johnson T, Gerstein M (2002). "Molecular fossils in the human genome: identification and analysis of the pseudogenes in chromosomes 21 and 22". Genome Res 12 (2): 272–80.

Hayashi G, Hagihara M, Nakatani K (2005). "Application of L-DNA as a molecular tag". Nucleic Acids Symp Ser (Oxf) 49: 261–262.

Hershey A, Chase M (1952). "Independent functions of viral protein and nucleic acid in growth of bacteriophage". J Gen Physiol 36 (1): 39–56.

Houck ML, Kumamoto AT, Gallagher DS, Benirschke K (2001). "Comparative cytogenetics of the African elephant (Loxodonta africana) and Asiatic elephant (Elephas maximus)". Cytogenet. Cell Genet. 93 (3-4): 249–52.

Houdebine L. "Transgenic animal models in biomedical research". Methods Mol Biol 360: 163–202.

Hubscher U, Maga G, Spadari S (2002). "Eukaryotic DNA polymerases". Annu Rev Biochem 71: 133–63.

Hüttenhofer A, Schattner P, Polacek N (2005). "Non-coding RNAs: hope or hype?". Trends Genet 21 (5): 289–97.

Iftode C, Daniely Y, Borowiec J (1999). "Replication protein A (RPA): the eukaryotic SSB". Crit Rev Biochem Mol Biol 34 (3): 141–80.

Ionov Y, Peinado MA, Malkhosyan S, Shibata D, Perucho M (1993). "Ubiquitous somatic mutations in simple repeated sequences reveal a new mechanism for colonic carcinogenesis". Nature 363 (6429): 558-61.

Isaksson J, Acharya S, Barman J, Cheruku P, Chattopadhyaya J (2004). "Single-stranded adenine-rich DNA and RNA retain structural characteristics of their respective double-stranded conformations and show directional differences in stacking pattern". Biochemistry 43 (51): 15996–6010.

Jeffrey A (1985). "DNA modification by chemical carcinogens". Pharmacol Ther 28 (2): 237–72.

Jeffreys A, Wilson V., Thein S. "Individual-specific 'fingerprints' of human DNA.". Nature 316 (6023): 76–9.

Jenuwein T, Allis C (2001). "Translating the histone code". Science 293 (5532): 1074–80.

Johnson A, O'Donnell M (2005). "Cellular DNA replicases: components and dynamics at the replication fork". Annu Rev Biochem 74: 283–315.

Johnson Z, Chisholm S (2004). "Properties of overlapping genes are conserved across microbial genomes". Genome Res 14 (11): 2268–72.

Joyce C, Steitz T (1995). "Polymerase structures and function: variations on a theme?". J Bacteriol 177 (22): 6321–9.

Jukes TH, Osawa S, The genetic code in mitochondria and chloroplasts., Experientia. 1990 Dec 1;46(11-12):1117-26.

Kleiman, Yaakov. "The Cohanim/DNA Connection: The fascinating story of how DNA studies confirm an ancient biblical tradition". aish.com (January 13, 2000). Accessed 4 March 2006.

Klose R, Bird A (2006). "Genomic DNA methylation: the mark and its mediators". Trends Biochem Sci 31 (2): 89–97.

Knight, R.D. and Landweber, L.F. (1998). Rhyme or reason: RNA-arginine interactions and the genetic code. Chemistry & Biology 5(9), R215-R220. PDF version of manuscript

Knight, R.D.; Freeland S. J. and Landweber, L.F. (1999) The 3 Faces of the Genetic Code. Trends in the Biochemical Sciences 24(6), 241-247.

Lamb R, Horvath C (1991). "Diversity of coding strategies in influenza viruses". Trends Genet 7 (8): 261–6.

Leroi A. 2003. Mutants: On the form, varieties & errors of the human body. 1:16-17. Harper Collins 2003.

Leslie AG, Arnott S, Chandrasekaran R, Ratliff RL (1980). "Polymorphism of DNA double helices". J. Mol. Biol. 143 (1): 49–72.

Levene P, (1919). "The structure of yeast nucleic acid". J Biol Chem 40 (2): 415–24.

Lost Tribes of Israel, NOVA, PBS airdate: 22 February 2000. Transcript available from PBS.org, (last accessed on 4 March 2006)

Lu XJ, Shakked Z, Olson WK (2000). "A-form conformational motifs in ligand-bound DNA structures". J. Mol. Biol. 300 (4): 819-40.

Luger K, Mäder A, Richmond R, Sargent D, Richmond T (1997). "Crystal structure of the nucleosome core particle at 2.8 A resolution". Nature 389 (6648): 251–60.

Makalowska I, Lin C, Makalowski W (2005). "Overlapping genes in vertebrate genomes". Comput Biol Chem 29 (1): 1–12.

Mandelkern M, Elias J, Eden D, Crothers D (1981). "The dimensions of DNA in solution". J Mol Biol 152 (1): 153–61.

Meselson M, Stahl F (1958). "The replication of DNA in Escherichia coli". Proc Natl Acad Sci U S A 44 (7): 671–82.

Molecular Structure of Deoxypentose Nucleic Acids. Wilkins M.H.F., A.R. Stokes A.R. & Wilson, H.R. Nature 171, 738–740 (1953)Nature Archives (PDF)

Mount DM (2004). Bioinformatics: Sequence and Genome Analysis, 2, Cold Spring Harbor Laboratory Press.

Munroe S (2004). "Diversity of antisense regulation in eukaryotes: multiple mechanisms, emerging patterns". J Cell Biochem 93 (4): 664–71.

NCBI: "The Genetic Codes", Compiled by Andrzej (Anjay) Elzanowski and Jim Ostell

Neale MJ, Keeney S (2006). "Clarifying the mechanics of DNA strand exchange in meiotic recombination". Nature 442 (7099): 153-8.

Nugent C, Lundblad V (1998). "The telomerase reverse transcriptase: components and regulation". Genes Dev 12 (8): 1073–85.

O'Driscoll M, Jeggo P (2006). "The role of double-strand break repair - insights from human genetics". Nat Rev Genet 7 (1): 45–54.

Oh D, Kim Y, Rich A (2002). "Z-DNA-binding proteins can act as potent effectors of gene expression in vivo". Proc. Natl. Acad. Sci. U.S.A. 99 (26): 16666-71.

Pabo C, Sauer R (1984). "Protein-DNA recognition". Annu Rev Biochem 53: 293–321.

Pabo C, Sauer R (1984). "Protein-DNA recognition". Annu Rev Biochem 53: 293–321.

Palecek E (1991). "Local supercoil-stabilized DNA structures". Critical Reviews in Biochemistry and Molecular Biology 26 (2): 151–226.

Parker J (2003). "Computing with DNA.". EMBO Rep 4 (1): 7–10.

Parkinson G, Lee M, Neidle S (2002). "Crystal structure of parallel quadruplexes from human telomeric DNA". Nature 417 (6891): 876–80.

Pidoux A, Allshire R (2005). "The role of heterochromatin in centromere function". Philos Trans R Soc Lond B Biol Sci 360 (1455): 569–79.

Ponnuswamy P, Gromiha M (1994). "On the conformational stability of oligonucleotide duplexes and tRNA molecules". J Theor Biol 169 (4): 419–32.

Ratel D, Ravanat J, Berger F, Wion D (2006). "N6-methyladenine: the other methylated base of DNA". Bioessays 28 (3): 309–15.

Ronneberg T.A.; Landweber L.F. and Freeland S.J. (2000) Testing a biosynthetic theory of the genetic code: Fact or artifact? Proceedings of the National Academy of Sciences, USA 97(25), 13690-13695.

Rothenburg S, Koch-Nolte F, Haag F. "DNA methylation and Z-DNA formation as mediators of quantitative differences in the expression of alleles". Immunol Rev 184: 286–98.

Sakamura, T. (1918), Kurze Mitteilung uber die Chromosomenzahlen und die Verwandtschaftsverhaltnisse der Triticum-Arten. Bot. Mag., 32: 151-154.

Sandman K, Pereira S, Reeve J (1998). "Diversity of prokaryotic chromosomal proteins and the origin of the nucleosome". Cell Mol Life Sci 54 (12): 1350–64.

Schoeffler A, Berger J (2005). "Recent advances in understanding structure-function relationships in the type II topoisomerase mechanism". Biochem Soc Trans 33 (Pt 6): 1465–70.

Shigenaga M, Gimeno C, Ames B (1989). "Urinary 8-hydroxy-22 -deoxyguanosine as a biological marker of in vivo oxidative DNA damage". Proc Natl Acad Sci U S A 86 (24): 9697–701.

Spiegelman B, Heinrich R (2004). "Biological control through regulated transcriptional coactivators". Cell 119 (2): 157-67.

Stephens T, Bunde C, Fillmore B (2000). "Mechanism of action in thalidomide teratogenesis". Biochem Pharmacol 59 (12): 1489–99.

Szathmáry E (1992). "What is the optimum size for the genetic alphabet?". Proc Natl Acad Sci U S A 89 (7): 2614–8.

Takahashi I, Marmur J. (1963). "Replacement of thymidylic acid by deoxyuridylic acid in the deoxyribonucleic acid of a transducing phage for Bacillus subtilis". Nature 197: 794–5.

Tarrago-Litvak L, Andréola M, Nevinsky G, Sarih-Cottin L, Litvak S (1994). "The reverse transcriptase of HIV-1: from enzymology to therapeutic intervention". FASEB J 8 (8): 497–503.

Thanbichler M, Wang S, Shapiro L (2005). "The bacterial nucleoid: a highly organized and dynamic structure". J Cell Biochem 96 (3): 506–21.

The codon AUG both codes for methionine and serves as an initiation site: the first AUG in an mRNA's coding region is where translation into protein begins.

The Nobel Prize in Physiology or Medicine 1962 Nobelprize .org Accessed 22 Dec 06

Thomas J (2001). "HMG1 and 2: architectural DNA-binding proteins". Biochem Soc Trans 29 (Pt 4): 395–401.

Tuteja N, Tuteja R (2004). "Unraveling DNA helicases. Motif, structure, mechanism and function". Eur J Biochem 271 (10): 1849–63.

Umeko Semba, Yasuko Umeda, Yoko Shibuya, Hiroaki Okabe, Sumio Tanase and Tetsuro Yamamoto (2004). "Primary structures of guinea pig high- and low-molecular-weight kininogens". International Immunopharmacology 4 (10-11): 1391-1400.

Valerie K, Povirk L (2003). "Regulation and mechanisms of mammalian double-strand break repair". Oncogene 22 (37): 5792–812.

Vargason JM, Eichman BF, Ho PS (2000). "The extended and eccentric E-DNA structure induced by cytosine methylation or bromination". Nature Structural Biology 7: 758–761.

Venter J, et al. (2001). "The sequence of the human genome". Science 291 (5507): 1304–51.

Vispé S, Defais M (1997). "Mammalian Rad51 protein: a RecA homologue with pleiotropic functions". Biochimie 79 (9-10): 587-92.

Wahl M, Sundaralingam M (1997). "Crystal structures of A-DNA duplexes". Biopolymers 44 (1): 45–63.

Walsh C, Xu G. "Cytosine methylation and DNA repair". Curr Top Microbiol Immunol 301: 283–315.

Wang G, Vasquez KM (2006). "Non-B DNA structure-induced genetic instability". Mutat Res 598 (1–2): 103–119.

Wang J (2002). "Cellular roles of DNA topoisomerases: a molecular perspective". Nat Rev Mol Cell Biol 3 (6): 430–40.

Watson J, Crick F (1953). "Molecular structure of nucleic acids; a structure for deoxyribose nucleic acid". Nature 171 (4356): 737–8.

3

Overview of Major Genetic Research Organizations

MOLECULAR GENETICS

Molecular genetics is the field of biology which studies the structure and function of genes at a molecular level. The field studies how the genes are transferred from generation to generation. Molecular genetics employs the methods of genetics and molecular biology. It is so-called to differentiate it from other sub fields of genetics such as ecological genetics and population genetics. An important area within molecular genetics is the use of molecular information to determine the patterns of descent, and therefore the correct scientific classification of organisms: this is called molecular systematics.

Along with determining the pattern of descendants, molecular genetics helps in understanding genetic mutations that can cause certain types of diseases. Through utilizing the methods of genetics and molecular biology, molecular genetics discovers the reasons why traits are carried on and how and why some may mutate.

Forward Genetics

One of the first tools available to molecular geneticists is the forward genetic screen. The aim of this technique is to identify mutations that produce a certain phenotype. A mutagen is very often used to accelerate this process. Once mutants have been isolated, the mutated gene can be molecularly identified.

Reverse Genetics

While forward genetic screens are productive, a more straightforward approach would be to determine the phenotype that results from mutating a given gene. This is called reverse genetics. In some organisms, such as yeast and mice, it is possible to induce the deletion of a particular gene, creating a gene knockout. Alternatives include the random induction of DNA deletions and subsequent selection for deletions in a gene of interest, the application of RNA interference and the creation of transgenic organisms that do not express a gene of interest.

Gene Therapy

Many mutations in a person's genes result in severe medical condition. These mutations cause the protein encoded by that gene to malfunction and cells that rely on this protein cannot function properly. This can cause problems for the tissues or organs. Conditions related to gene mutations are called genetic disorders. One way to fix the internal problems is gene therapy. By adding a corrected copy of the gene, the affected cells, tissues, and organs may work properly. As opposed to drug-based approaches, gene therapy repairs the underlying genetic defect.

Gene Therapy is the process to treat or alleviate diseases by genetically modifying the cells of the infected person, causing the gene to function properly. When a human disease gene has been recognized, molecular genetics tools can be used to explore the process of the gene in both the normal and pathogenic states. From there, the gene is transferred either in vivo or ex vivo and the body begins to identify the new gene. Gene therapy has to be repeated several times for the infected patient to continually be relieved.

Currently, gene therapy is still being experimented with and products are not approved by the U.S. Food and Drug Administration. There have been several set backs in the last 15 years that have restricted further measures on gene therapy. As there are unsuccessful attempts, there continues to be a growing number of successful gene therapy transfers which has furthered the research.

Major diseases that can be treated with gene therapy: infectious diseases (result of infection by a virus or bacterial pathogen), cancers, inherited disorders, immune system disorders, etc.

Classical Gene Therapy

Classical gene therapy is the approach which delivers genes to the appropriate target cells with a goal of attaining optimal expression of the new, introduced gene. Once inside the patient, the expressed genes are intended to: produce a product that the patient lacks, kill the diseased cells directly by producing a toxin, and activate the immune system to help the killing of the diseased cells.

Nonclassical Gene Therapy

Nonclassical gene therapy inhibits the expression of genes related to pathogenesis, or corrects a genetic defect and restores normal gene expression.

In Vivo Gene Transfer

During In vivo gene transfer the genes are transferred directly into the tissue of the patient. This can be the only possible option in patients with tissues where individual cells cannot be cultured outside the body in sufficient numbers (i.e. brain cells). Also, in vivo gene transfer is necessary when cultured cells cannot be re-implanted in patients effectively.

Ex Vivo Gene Transfer

During Ex vivo gene transfer the cells are cultured outside the body and then the genes are transferred into the cells grown in culture. The cells that have been transformed successfully are expanded by cell culture and then introduced into the patient.

Principles for Gene Transfer

Classical gene therapies usually require efficient transfer of cloned genes into the disease cells to the introduced genes are expressed at high levels. There are several different physicochemical and biological methods that can be used to transfer genes into human cells. The size of the DNA fragments that can be transferred are very limited, and often the transferred gene is not a conventional gene. Horizontal gene transfer is the transfer of genetic material from one cell to another that is not its offspring. Artificial horizontal gene transfer is a form of genetic engineering.

Techniques in Molecular Genetics

There are three general techniques used for molecular genetics: amplification, separation and detection, and expression. Specifically used for amplification is the Polymerase chain reaction, which is an "indispensable tool in a great variety of applications". In the separation and detection technique DNA and mRNA are isolated from their cells. Gene expression in cells or organisms is done in a place or time that is not normal for that specific gene.

Amplification

There are other methods for amplification besides the Polymerase Chain Reaction. Cloning DNA in bacteria is also a way to amplify DNA in genes.

Polymerase Chain Reaction

The main materials used in the Polymerase Chain Reaction are DNA nucleotides, template DNA, primers and Taq polymerase. DNA nucleotides are the base for the new DNA, the template DNA is the specific sequence being amplified, primers are complementary nucleotides that can go on either side of the template DNA, and Taq polymerase is a heat stable enzyme that jump-starts the production of new DNA.

This technique does not need to use living bacteria or cells, all that is needed is the base sequence of the DNA needing amplification.

Cloning DNA in Bacteria

The word cloning for this type of amplification entails making multiple identical copies of a sequence of DNA. The target DNA sequence is then inserted into a [cloning vector]. Because this vector originates from a self-replicating virus, plasmid, or higher organism cell when the appropriate size DNA is inserted the "target and vector DNA fragments are then ligated" and create a recombinant DNA molecule. The recombinant DNA molecules are then put into a bacteria strain (usually *E. coli*) which produces several identical copies by transformation. Transformation is the DNA uptake mechanism possessed by bacteria. However, only one recombinant DNA molecule can be cloned within a single bacteria cell, so each clone is of just one DNA insert.

Separation and Detection

In separation and detection DNA and mRNA are isolated from cells (the separation) and then detected simply by the isolation. Cell cultures are also grown to provide a constant supply of cells ready for isolation.

Cell Cultures

A cell culture for molecular genetics is a culture that is grown in artificial conditions. Some cell types grow well in cultures such a skin cells, but other cells are not as productive in cultures. There are different techniques for each type of cell, some only recently being found to foster growth in stem and nerve cells. Cultures for molecular genetics are frozen in order to preserve all copies of the gene specimen and thawed only when needed. This allows for a steady supply of cells.

DNA Isolation

DNA isolation extracts DNA from a cell in a pure form. First, the DNA is separated from cellular components such as proteins, RNA, and lipids. This is done by placing the chosen cells in a tube with a solution that mechanically, chemically, breaks the cells open. This solution contains enzymes, chemicals, and salts that breaks down the cells except for the DNA. It contains enzymes to dissolve proteins, chemicals to destroy all RNA present, and salts to help pull DNA out of the solution.

Next, the DNA is separated from the solution by being spun in a centrifuge, which allows the DNA to collect in the bottom of the tube. After this cycle in the centrifuge the solution is poured off and the DNA is resuspended in a second solution that makes the DNA easy to work with in the future.

This results in a concentrated DNA sample that contains thousands of copies of each gene. For large scale projects such as sequencing the human genome, all this work is done by robots.

mRNA Isolation

Expressed DNA that codes for the synthesis of a protein is the final goal for scientists and this expressed DNA is obtained by isolation mRNA (Messenger RNA). First, laboratories use a normal cellular modification of mRNA that adds up to 200 adenine nucleotides to the end of the molecule (poly(A) tail). Once this has been added, the cell is ruptured and its cell contents are exposed to synthetic beads that are coated with thymine string nucleotides. Because Adenine and Thymine pair together in DNA, the poly(A) tail and synthetic beads are attracted to one another, and once they bind in this process the cell components can be washed away without removing the mRNA. Once the mRNA has been isolated, reverse transcriptase is employed to convert it to single-stranded DNA, from which a stable double-stranded DNA is produced using DNA polymerase. Complementary DNA (cDNA) is much more stable than mRNA and so, once the double-stranded DNA has been produced it represents the expressed DNA sequence scientists look for.

The Molecular Genetics Project

The Human Genome Project is a molecular genetics project that began in 1990 and was projected to take fifteen years to complete. However, because of technological advances the progress of the project was advanced and the project finished in 2003, taking only thirteen years. The project was started by the U.S. Department of Energy and the National Institutes of Health in an effort to reach six set goals. These goals included:

1. identifying 20,000 to 25,000 genes in human DNA (although initial estimate were approximately 100,000 genes),
2. determining sequences of chemical based pairs in human DNA,
3. storing all found information into databases,
4. improving the tools used for data analysis,
5. transferring technologies to private sectors, and
6. addressing the ethical, legal, and social issues (ELSI) that may arise from the projects.

The project was worked on by eighteen different countries including the United States, Japan, France, Germany, and the United Kingdom. The collaborative effort resulted in the discovery of the many benefits of molecular genetics. Discoveries such as molecular medicine, new energy sources and environmental applications,

DNA forensics, and livestock breeding, are only a few of the benefits that molecular genetics can provide.

POPULATION GENETICS

Population genetics is the study of the allele frequency distribution and change under the influence of the four evolutionary forces: natural selection, genetic drift, mutation and gene flow. It also takes account of population subdivision and population structure in space. As such, it attempts to explain such phenomena as adaptation and speciation. Population genetics was a vital ingredient in the modern evolutionary synthesis, its primary founders were Sewall Wright, J. B. S. Haldane and R. A. Fisher, who also laid the foundations for the related discipline of quantitative genetics.

Scope and Theoretical Considerations

Perhaps the most significant "formal" achievement of the modern evolutionary synthesis has been the framework of mathematical population genetics. Indeed some authors (Beatty 1986) would argue that it does *define* the core of the modern synthesis.

Lewontin (1974) outlined the theoretical task for population genetics. He imagined two spaces: a "genotypic space" and a "phenotypic space". The challenge of a *complete* theory of population genetics is to provide a set of laws that predictably map a population of genotypes (G_1) to a phenotype space (P_1), where selection takes place, and another set of laws that map the resulting population (P_2) back to genotype space (G_2) where Mendelian genetics can predict the next generation of genotypes, thus completing the cycle. Even leaving aside for the moment the non-Mendelian aspects revealed by molecular genetics, this is clearly a gargantuan task. Visualizing this transformation schematically:

$$G_1 \rightarrow^{T_2} P_1 \rightarrow^{T_2} P_2 \rightarrow^{T_3} G_2 \rightarrow^{T_4} G_2 \rightarrow \ldots$$

(adapted from Lewontin 1974, p. 12).

T^1 represents the genetic and epigenetic laws, the aspects of functional biology, or development, that transform a genotype into phenotype. We will refer to this as the "genotype-phenotype map". T^2 is the transformation due to natural selection, T^3 are epigenetic relations that predict genotypes based on the selected phenotypes and finally T^4 the rules of Mendelian genetics.

In practice, there are two bodies of evolutionary theory that exist in parallel, traditional population genetics operating in the genotype space and the biometric theory used in plant and animal breeding, operating in phenotype space. The missing part is the mapping between the genotype and phenotype space. This leads to a "sleight of hand" (as Lewontin terms it) whereby variables in the equations of one domain, are considered parameters or *constants*, where, in a full-treatment they would be transformed themselves by the evolutionary process and are in reality *functions* of the state variables in the other domain. The "sleight of hand" is assuming that we know this mapping. Proceeding as if we do understand it is enough to analyze many cases of interest. For example, if the phenotype is almost one-to-one with genotype (sickle-cell disease) or the time-scale is sufficiently short, the "constants" can be treated as such; however, there are many situations where it is inaccurate.

Population Geneticists

The three founders of population genetics were the Britons R.A. Fisher and J.B.S. Haldane and the American Sewall Wright. Fisher and Wright had some fundamental disagreements and a controversy about the relative roles of selection and drift continued for much of the century between the Americans and the British. The Frenchman Gustave Malécot was also important early in the development of the discipline. John Maynard Smith was Haldane's pupil, whilst W.D. Hamilton was heavily influenced by the writings of Fisher. The American George R. Price worked with both Hamilton and Maynard Smith. On the American side, Richard Lewontin and the Japanese Motoo Kimura were heavily influenced by Wright.

QUANTITATIVE GENETICS

Quantitative genetics is the study of continuous traits (such as height or weight) and its underlying mechanisms. It is effectively an extension of simple Mendelian inheritance in that the combined effect of the many underlying genes results in a continuous distribution of phenotypic values.

History

The field was founded, in evolutionary terms, by the originators of the modern synthesis, R.A. Fisher, Sewall Wright and J.B.S. Haldane, and aimed to predict the response to selection given data on the phenotype and relationships of individuals.

Analysis of Quantitative trait loci, or QTL, is a more recent addition to the study of quantitative genetics. A QTL is a region in the genome that affects the trait or traits of interest. Quantitative trait loci approaches require accurate phenotypic, pedigree and genotypic data from a large number of individuals.

Traits

Quantitative genetics is not limited to continuous traits, but to all traits that are determined by many genes. This includes:

- Continuous traits are quantitative traits with a continuous phenotypic range. They are often polygenic, and may also be influenced significantly by environmental effects.
- Meristic traits or other ordinal numbers are expressed in whole numbers, such as number of offspring, or number of bristles on a fruit fly. These traits can be either treated as approximately continuous traits or as threshold traits.
- Some qualitative traits can be treated as if they have an underlying quantitative basis, expressed as a threshold trait (or multiple thresholds). Some human diseases (such as, schizophrenia) have been studied in this manner.

CLASSICAL GENETICS

Classical genetics consists of the techniques and methodologies of genetics that predate the advent of molecular biology. A key discovery of classical genetics in eukaryotes was genetic linkage. The observation that some genes do not segregate independently at meiosis, broke the laws of Mendelian inheritance, and provided science with a way to map characteristics to a location on the chromosomes. Linkage maps are still used today, especially in breeding for plant improvement.

After the discovery of the genetic code and such tools of cloning as restriction enzymes, the avenues of investigation open to geneticists were greatly broadened. Some classical genetic ideas have been supplanted with the mechanistic understanding brought by molecular discoveries, but many remain intact and in use. Classical genetics is often contrasted with reverse genetics, and aspects of molecular biology are sometimes referred to as molecular genetics.

ECOLOGICAL GENETICS

Ecological genetics is the study of genetics in the context of the interactions among organisms and between the organisms and their environment. While molecular genetics studies the structure and function of genes at a molecular level, ecological genetics (and the related field of population genetics) studies phenotypic evolution in natural populations of organisms. Research in this field is of traits of ecological significance — that is, traits related to fitness, which affect an organism's survival and reproduction (e.g., flowering time, drought tolerance, sex ratio).

Studies are often done on insects and other organisms that have short generation times, and thus evolve at fast rates.

History

Although work on natural populations had been done previously, it is acknowledged that the field was founded by the English biologist E.B. Ford (1901-1988) in the early 20th century. Ford was taught genetics at Oxford University by Julian Huxley, and started research on the genetics of natural populations in 1924. *Ecological Genetics* is the title of his 1964 'magnum opus' on the subject (4th ed 1975). Other notable ecological geneticists would include Theodosius Dobzhansky who worked on chromosome polymorphism in fruit flies. As a young researcher in Russia, Dobzhansky had been influenced by Sergei Chetverikov, who also deserves to be remembered as a founder of genetics in the field, though his significance was not appreciated until much later.

Philip Sheppard, Cyril Clarke, Bernard Kettlewell and A.J. Cain were all strongly influenced by Ford; their careers date from the post WWII era. Collectively, their work on lepidoptera, and on human blood groups, established the field, and threw light on selection in natural populations where its role had been once doubted.

Work of this kind needs long-term funding, as well as grounding in both ecology and genetics. These are both difficult requirements. Research projects can last longer than a researcher's career; for instance, research into mimicry started 150 years ago, and is still going strongly. Funding of this type of research is still rather erratic, but at least the value of working with natural populations in the field cannot now be doubted.

GENE

A gene is a locatable region of genomic sequence, corresponding to a unit of inheritance, which is associated with regulatory regions, transcribed regions and/or other functional sequence regions. The physical development and phenotype of organisms can be thought of as a product of genes interacting with each other and with the environment. A concise definition of gene taking into account complex patterns of regulation and transcription, genic conservation and non-coding RNA genes, has been proposed by Gerstein et al. "A gene is a union of genomic sequences encoding a coherent set of potentially overlapping functional products".

In cells, genes consist of a long strand of DNA that contains a promoter, which controls the activity of a gene, and coding and non-coding sequence. Coding sequence determines what the gene produces, while non-coding sequence can regulate the conditions of gene expression. When a gene is active, the coding and non-coding sequence is copied in a process called transcription, producing an RNA copy of the gene's information. This RNA can then direct the synthesis of proteins via the genetic code. However, RNAs can also be used directly, for example as part of the ribosome. These molecules resulting from gene expression, whether RNA or protein, are known as gene products.

Genes often contain regions that do not encode products, but regulate gene expression. The genes of eukaryotic organisms can contain regions called introns that are removed from the messenger RNA in a process called splicing. The regions encoding gene products are called exons. In eukaryotes, a single gene can encode multiple proteins, which are produced through the creation of different arrangements of exons through *alternative splicing*. In prokaryotes (bacteria and archaea), introns are less common and genes often contain a single uninterrupted stretch of DNA, called a *cistron*, that codes for a product. Prokaryotic genes are often arranged in groups called operons with promoter and operator sequences that regulate transcription of a single long RNA. This RNA contains multiple coding sequences. Each coding sequence is preceded by a Shine-Dalgarno sequence that ribosomes recognize.

The total set of genes in an organism is known as its genome. An organism's genome size is generally lower in prokaryotes, both in number of base pairs and number of genes, than even single-celled eukaryotes. However, there is no clear relationship between genome sizes and complexity in eukaryotic organisms. One of the largest known genomes belongs to the single-celled amoeba *Amoeba dubia*, with over 670 billion base pairs, some 200 times larger than the human genome. The estimated number of genes in the human genome has been repeatedly revised downward since the completion of the Human Genome Project; current estimates place the human genome at just under 3 billion base pairs and about 20,000–25,000 genes.. A recent *Science* article gives a final number of 20,488, with perhaps 100 more yet to be discovered. The gene density of a genome is a measure of the number of genes per million base pairs (called a megabase, Mb); prokaryotic genomes have much higher gene densities than eukaryotes. The gene density of the human genome is roughly 12–15 genes/Mb.

History

The existence of genes was first suggested by Gregor Mendel (1822-1884), who, in the 1860s, studied inheritance in pea plants and hypothesized a factor that conveys traits from parent to offspring. He spent over 10 years of his life on one experiment. Although he did not use the term *gene*, he explained his results in terms of inherited characteristics. Mendel was also the first to hypothesize independent assortment, the distinction between dominant and recessive traits, the distinction between a heterozygote and homozygote, and the difference between what would later be described as genotype and phenotype. Mendel's concept was given a name by Hugo de Vries in 1889, who, at that time probably unaware of Mendel's work, in his book *Intracellular Pangenesis* coined the term "pangen" for "the smallest particle [representing] one hereditary characteristic". Wilhelm Johannsen abbreviated this term to "gene" ("gen" in Danish and German) two decades later.

In the early 1900s, Mendel's work received renewed attention from scientists. In 1910, Thomas Hunt Morgan showed that genes reside on specific chromosomes. He later showed that genes occupy specific locations on the chromosome. With this knowledge, Morgan and his students began the first chromosomal map of the fruit fly *Drosophila*. In 1928, Frederick Griffith showed that genes could be transferred. In what is now known as Griffith's experiment, injections into a mouse of a deadly strain of bacteria that had been heat-killed transferred genetic information to a safe strain of the same bacteria, killing the mouse.

In 1941, George Wells Beadle and Edward Lawrie Tatum showed that mutations in genes caused errors in certain steps in metabolic pathways. This showed that specific genes code for specific proteins, leading to the "one gene, one enzyme" hypothesis. Oswald Avery, Collin Macleod, and Maclyn McCarty showed in 1944 that DNA holds the gene's information. In 1953, James D. Watson and Francis Crick demonstrated the molecular structure of DNA. Together, these discoveries established the central dogma of molecular biology, which states that proteins are translated from RNA which is transcribed from DNA. This dogma has since been shown to have exceptions, such as reverse transcription in retroviruses.

In 1972, Walter Fiers and his team at the Laboratory of Molecular Biology of the University of Ghent (Ghent, Belgium) were the first to determine the sequence of a gene: the gene for Bacteriophage MS2 coat protein. Richard J. Roberts and Phillip Sharp discovered in 1977 that genes can be split into segments. This leads to the idea that one gene can make several proteins. Recently (as of 2003-2006), biological results let the notion of gene appear more slippery. In particular, genes do not seem to sit side by side on DNA like discrete beads. Instead, regions of the DNA producing distinct proteins may overlap, so that the idea emerges that "genes are one long continuum".

Mendelian Inheritance and Classical Genetics

Darwin used the term Gemmule to describe a microscopic unit of inheritance, and what would later become known as Chromosomes had been observed separating out during cell division by Wilhelm Hofmeister as early as 1848. The idea that chromosomes were the carriers of inheritance was expressed in 1883 by Wilhelm Roux. The modern conception of the gene originated with work by Gregor Mendel, a 19th century Augustinian monk who systematically studied heredity in pea plants. Mendel's work was the first to illustrate particulate inheritance, or the theory that inherited traits are passed from one generation to the next in discrete units that interact in well-defined ways. Danish botanist Wilhelm Johannsen coined the word "gene" in 1909 to describe these fundamental physical and functional units of heredity, while the related word genetics was first used by William Bateson in 1905.

The word was derived from Hugo De Vries' 1889 term *pangen* for the same concept, itself a derivative of the word *pangenesis* coined by Darwin (1868). The word pangenesis is made from the Greek words *pan* (a prefix meaning "whole", "encompassing") and *genesis* ("birth") or *genos* ("origin").

According to the theory of Mendelian inheritance, variations in phenotype - the observable physical and behavioral characteristics of an organism - are due to variations in genotype, or the organism's particular set of genes, each of which specifies a particular trait. Different genes for the same trait, which give rise to different phenotypes, are known as alleles. Organisms such as the pea plants Mendel worked on, along with many plants and animals, have two alleles for each trait, one inherited from each parent. Alleles may be dominant or recessive; dominant alleles give rise to their corresponding phenotypes when paired with any other allele for the same trait, while recessive alleles give rise to their corresponding phenotype only when paired with another copy of the same allele. For example, if the allele specifying tall stems in pea plants is dominant over the allele specifying short stems, then pea plants that inherit one tall allele from one parent and one short allele from the other parent will also have tall stems. Mendel's work found that alleles assort independently in the production of gametes, or germ cells, ensuring variation in the next generation.

Prior to Mendel's work, the dominant theory of heredity was one of blending inheritance, which proposes that the traits of the parents blend or mix in a smooth, continuous gradient in the offspring. Although Mendel's work was largely unrecognized after its first publication in 1866, it was rediscovered in 1900 by three European scientists, Hugo de Vries, Carl Correns, and Erich von Tschermak, who had reached similar conclusions from their own research. However, these scientists were not yet aware of the identity of the 'discrete units' on which genetic material resides.

A series of subsequent discoveries led to the realization decades later that chromosomes within cells are the carriers of genetic material, and that they are made of DNA (deoxyribonucleic acid), a polymeric molecule found in all cells on which the 'discrete units' of Mendelian inheritance are encoded. The modern study of genetics at the level of DNA is known as molecular genetics and the synthesis of molecular genetics with traditional Darwinian evolution is known as the modern evolutionary synthesis.

Physical Definitions

The vast majority of living organisms encode their genes in long strands of DNA. DNA consists of a chain made from four types of nucleotide subunits: adenosine, cytidine, guanosine, and thymidine. Each nucleotide subunit consists of three components: a phosphate group, a deoxyribose sugar ring, and a nucleobase. Thus, nucleotides in DNA or RNA are typically called 'bases'; consequently they are commonly referred to simply by their purine or pyrimidine original base components adenine, cytosine, guanine, thymine. Adenine and guanine are purines and cytosine and thymine are pyrimidines. The most common form of DNA in a cell is in a double helix structure, in which two individual DNA strands twist around each other in a right-handed spiral. In this structure, the base pairing rules specify that guanine pairs with cytosine and adenine pairs with thymine (each pair contains one purine and one pyrimidine). The base pairing between guanine and cytosine forms three hydrogen bonds, while the base pairing between adenine and thymine forms two hydrogen bonds. The two strands in a double helix must therefore be *complementary*, that is, their bases must align such that the adenines of one strand are paired with the thymines of the other strand, and so on.

Due to the chemical composition of the pentose residues of the bases, DNA strands have directionality. One end of a DNA polymer contains an exposed hydroxyl group on the deoxyribose, this is known as the 3' end of the molecule. The other end contains an exposed phosphate group, this is the 5' end. The directionality of DNA is vitally important to many cellular processes, since double helices are necessarily directional (a strand running 5'-3' pairs with a complementary strand running 3'-5') and processes such as DNA replication occur in only one direction. All nucleic acid synthesis in a cell occurs in the 5'-3' direction, because new monomers are added via a dehydration reaction that uses the exposed 3' hydroxyl as a nucleophile.

The expression of genes encoded in DNA begins by transcribing the gene into RNA, a second type of nucleic acid that is very similar to DNA, but whose monomers contain the sugar ribose rather than deoxyribose. RNA also contains the base uracil in place of thymine. RNA molecules are less stable than DNA and are typically single-stranded. Genes that encode proteins are composed of a series of three-nucleotide sequences called codons, which serve as the "words" in the genetic "language". The genetic code specifies the correspondence during protein translation between codons and amino acids. The genetic code is nearly the same for all known organisms.

RNA Genes

In some cases, RNA is an intermediate product in the process of manufacturing proteins from genes. However, for other gene sequences, the RNA molecules are the actual functional products. For example, RNAs known as ribozymes are capable of enzymatic function, and miRNAs have a regulatory role. The DNA sequences from which such RNAs are transcribed are known as non-coding DNA, or RNA genes.

Some viruses store their entire genomes in the form of RNA, and contain no DNA at all. Because they use RNA to store genes, their cellular hosts may synthesize their proteins as soon as they are infected and without the delay in waiting for transcription. On the other hand, RNA retroviruses, such as HIV, require the reverse transcription of their genome from RNA into DNA before their proteins can be synthesized. In 2006, French researchers came across a puzzling example of RNA-mediated inheritance in mouse. Mice with a loss-of-function mutation in the gene Kit have white tails. Offspring of these mutants can have white tails despite having only normal Kit genes. The research team traced this effect back to mutated Kit RNA. While RNA is common as genetic storage material in viruses, in mammals in particular RNA inheritance has been observed very rarely.

Functional Structure of a Gene

All genes have regulatory regions in addition to regions that explicitly code for a protein or RNA product. A universal regulatory region shared by all genes is known as the promoter, which provides a position that is recognized by the transcription machinery when a gene is about to be transcribed and expressed. Although promoter regions have a consensus sequence that is the most common sequence at this position, some genes have "strong" promoters that bind the transcription machinery well, and others have "weak" promoters that bind poorly. These weak promoters usually permit a lower rate of transcription than the strong promoters, because the transcription machinery binds to them and initiates transcription less frequently. Other possible regulatory regions include enhancers, which can compensate for a weak promoter. Most regulatory regions are "upstream" — that is, before or toward the 5' end of the transcription initiation site. Eukaryotic promoter regions are much more complex and difficult to identify than prokaryotic promoters.

Many prokaryotic genes are organized into operons, or groups of genes whose products have related functions and which are transcribed as a unit. By contrast, eukaryotic genes are transcribed only one at a time, but may include long stretches of DNA called introns which are transcribed but never translated into protein (they are spliced out before translation). Splicing can also occur in prokaryotic genes, but is less common than in eukaryotes.

Chromosomes

The total complement of genes in an organism or cell is known as its genome, which may be stored on one or more chromosomes; the region of the chromosome at which a particular gene is located is called its locus. A chromosome consists of a single, very long DNA helix on which thousands of genes are encoded. Prokaryotes - bacteria and archaea - typically store their genomes on a single large, circular chromosome, sometimes supplemented by additional small circles of DNA called plasmids, which usually encode only a few genes and are easily transferable between individuals. For example, the genes for antibiotic resistance are usually encoded on bacterial plasmids and can be passed between individual cells, even those of different species, via horizontal gene transfer. Although some simple eukaryotes also possess plasmids with small numbers of genes, the majority of eukaryotic genes are stored on multiple linear chromosomes, which are packed within the nucleus in complex with storage proteins called histones. The manner in which DNA is stored on the histone, as well as chemical modifications of the histone itself, are regulatory mechanisms governing whether a particular region of DNA is accessible for gene expression. The ends of eukaryotic chromosomes are capped by long stretches of repetitive sequences called telomeres, which do not code for any gene product but are present to prevent degradation of coding and regulatory regions during DNA replication. The length of the telomeres tends to decrease each time the genome is replicated in preparation for cell division; the loss of telomeres has been proposed as an explanation for cellular senescence, or the loss of the ability to divide, and by extension for the aging process in organisms.

While the chromosomes of prokaryotes are relatively gene-dense, those of eukaryotes often contain so-called "junk DNA", or regions of DNA that serve no obvious function. Simple single-celled eukaryotes have relatively small amounts of such DNA, while the genomes of complex multicellular organisms, including humans, contain an absolute majority of DNA without an identified function. However it now appears that, although protein-coding DNA makes up barely 2% of the human genome, about 80% of the bases in the genome may be being expressed, so the term "junk DNA" may be a misnomer.

Gene Expression

In all organisms, there are two major steps separating a protein-coding gene from its protein: first, the DNA on which the gene resides must be *transcribed* from DNA to messenger RNA (mRNA), and second, it must be *translated* from mRNA to protein. RNA-coding genes must still go through the first step, but are not translated into protein. The process of producing a biologically functional molecule of either RNA or protein is called gene expression, and the resulting molecule itself is called a gene product.

Genetic Code

The genetic code is the set of rules by which a gene is translated into a functional protein. Each gene consists of a specific sequence of nucleotides encoded in a DNA (or sometimes RNA) strand; a correspondence between nucleotides, the basic building blocks of genetic material, and amino acids, the basic building blocks of proteins, must be established for genes to be successfully translated into functional proteins. Sets of three nucleotides, known as codons, each correspond to a specific amino acid or to a signal; three codons are known as "stop codons" and, instead of specifying a new amino acid, alert the translation machinery that the end of the gene has been reached. There are 64 possible codons (four possible nucleotides at each of three positions, hence 43 possible codons) and only 20 standard amino acids; hence the code is redundant and multiple codons can specify the same amino acid. The correspondence between codons and amino acids is nearly universal among all known living organisms.

Transcription

The process of genetic transcription produces a single-stranded RNA molecule known as messenger RNA, whose nucleotide sequence is complementary to the DNA from which it was transcribed. The DNA strand whose sequence matches that of the RNA is known as the coding strand and the strand from which the RNA was synthesized is the template strand. Transcription is performed by an enzyme called an RNA polymerase, which reads the template strand in the 3' to 5' direction and synthesizes the RNA from 5' to 3'. To initiate transcription,the polymerase first recognizes and binds a promoter region of the gene. Thus a major mechanism of gene regulation is the blocking or sequestering of the promoter region, either by tight binding by repressor molecules that physically block the polymerase, or by organizing the DNA so that the promoter region is not accessible.

In prokaryotes, transcription occurs in the cytoplasm; for very long transcripts, translation may begin at the 5' end of the RNA while the 3' end is still being transcribed. In eukaryotes, transcription necessarily occurs in the nucleus, where the cell's DNA is sequestered; the RNA molecule produced by the polymerase is known as the primary transcript and must undergo post-transcriptional modifications before being exported to the cytoplasm for translation. The splicing of introns present within the transcribed region is a modification unique to eukaryotes; alternative splicing mechanisms can result in mature transcripts from the same gene having different sequences and thus coding for different proteins. This is a major form of regulation in eukaryotic cells.

Translation

Translation is the process by which a mature mRNA molecule is used as a template for synthesizing a new protein. Translation is carried out by ribosomes, large complexes of RNA and protein responsible for carrying out the chemical reactions to add new amino acids to a growing polypeptide chain by the formation of peptide bonds. The genetic code is read three nucleotides at a time, in units called codons, via interactions with specialized RNA molecules called transfer RNA (tRNA). Each tRNA has three unpaired bases known as the anticodon that are complementary to the codon it reads; the tRNA is also covalently attached to the amino acid specified by the complementary codon. When the tRNA binds to its complementary codon in an mRNA strand, the ribosome ligates its amino acid cargo to the new polypeptide chain, which is synthesized from amino terminus to carboxyl terminus. During and after its synthesis, the new protein must fold to its active three-dimensional structure before it can carry out its cellular function.

DNA Replication and Inheritance

The growth, development, and reproduction of organisms relies on cell division, or the process by which a single cell divides into two usually identical daughter cells. This requires first making a duplicate copy of every gene in the genome in a process called DNA replication. The copies are made by specialized enzymes known as DNA polymerases, which "read" one strand of the double-helical DNA, known as the template strand, and synthesize a new complementary strand. Because the DNA double helix is held together by base pairing, the sequence of one strand completely specifies the sequence of its complement; hence only one strand needs to be read by the enzyme to produce a faithful copy. The process of DNA replication is semiconservative; that is, the copy of the genome inherited by each daughter cell contains one original and one newly synthesized strand of DNA.

After DNA replication is complete, the cell must physically separate the two copies of the genome and divide into two distinct membrane-bound cells. In prokaryotes - bacteria and archaea - this usually occurs via a relatively simple process called binary fission, in which each circular genome attaches to the cell membrane and is separated into the daughter cells as the membrane invaginates to split the cytoplasm into two membrane-bound portions. Binary fission is extremely fast compared to the rates of cell division in eukaryotes. Eukaryotic cell division is a more complex process known as the cell cycle; DNA replication occurs during a phase of this cycle known as S phase, while the process of segregating chromosomes and splitting the cytoplasm occurs during M phase. In many single-celled eukaryotes such as yeast, reproduction by budding is common, which results in asymmetrical portions of cytoplasm in the two daughter cells.

Molecular Inheritance

The duplication and transmission of genetic material from one generation of cells to the next is the basis for molecular inheritance, and the link between the classical and molecular pictures of genes. Organisms inherit the characteristics of their parents because the cells of the offspring contain copies of the genes in their parents' cells. In asexually reproducing organisms, the offspring will be a genetic copy or clone of the parent organism. In sexually reproducing organisms, a specialized form of cell division called meiosis produces cells called gametes or germ cells that are haploid, or contain only one copy of each gene. The gametes produced by females are called eggs or ova, and those produced by males are called sperm. Two gametes fuse to form a fertilized egg, a single cell that once again has a diploid number of genes - each with one copy from the mother and one copy from the father.

During the process of meiotic cell division, an event called genetic recombination or *crossing-over* can sometimes occur, in which a length of DNA on one chromatid is swapped with a length of DNA on the corresponding sister chromatid. This has no effect if the alleles on the chromatids are the same, but results in reassortment of otherwise linked alleles if they are different. The Mendelian principle of independent assortment asserts that each of a parent's two genes for each trait will sort independently into gametes; which allele an organism inherits for one trait is unrelated to which allele it inherits for another trait. This is in fact only true for genes that do not reside on the same chromosome, or are located very far from one another on the same chromosome. The closer two genes lie on the same chromosome, the more closely they will be associated in gametes and the more often they will appear together; genes that are very close are essentially never separated because it is extremely unlikely that a crossover point will occur between them. This is known as genetic linkage.

Mutation

DNA replication is for the most part extremely accurate, with an error rate per site of around 10-6 to 10-10 in eukaryotes. Rare, spontaneous alterations in the base sequence of a particular gene arise from a number of sources, such as errors in DNA replication and the aftermath of DNA damage. These errors are called mutations. The cell contains many DNA repair mechanisms for preventing mutations and maintaining the integrity of the genome; however, in some cases — such as breaks in both DNA strands of a chromosome — repairing the physical damage to the molecule is a higher priority than producing an exact copy. Due to the degeneracy of the genetic code, some mutations in protein-coding genes are *silent*, or produce no change in the amino acid sequence of the protein for which they code; for example, the codons UCU and UUC both code for serine, so the U"!C mutation has no effect on the protein. Mutations that do have phenotypic effects are most often neutral or deleterious to the organism, but sometimes they confer benefits to the organism's fitness.

Mutations propagated to the next generation lead to variations within a species' population. Variants of a single gene are known as alleles, and differences in alleles may give rise to differences in traits. Although it is rare for the variants in a single gene to have clearly distinguishable phenotypic effects, certain well-defined traits are in fact controlled by single genetic loci. A gene's most common allele is called the wild type allele, and rare alleles are called mutants. However, this does not imply that the wild-type allele is the ancestor from which the mutants are descended.

The Genome

Chromosomal Organization

The total complement of genes in an organism or cell is known as its genome. In prokaryotes, the vast majority of genes are located on a single chromosome of circular DNA, while eukaryotes usually possess multiple individual linear DNA helices packed into dense DNA-protein complexes called chromosomes. Extrachromosomal DNA is present in many prokaryotes and some simple eukaryotes as small, circular pieces of DNA called plasmids, which usually contain only a few genes each. Generally, regulatory regions and junk DNA are considered to be part of an organism's genome, but structural regions such as telomeres are not. The location (or locus) of a gene and the chromosome on which it is situated is in a sense arbitrary. Genes that appear together on the chromosomes of one species, such as humans, may appear on separate chromosomes in another species, such as mice. Two genes positioned near one another on a chromosome may encode proteins that figure in the same cellular process or in completely unrelated processes. As an example of the former, many of the genes involved in spermatogenesis reside together on the Y chromosome.

Many species carry more than one copy of their genome within each of their somatic cells. Cells or organisms with only one copy of each gene are called haploid; those with two copies are called diploid; and those with more than two copies are called polyploid. When more than one copy is present, the two copies are not necessarily identical; in sexually reproducing organisms, one copy is normally inherited from each parent. The copies may contain distinct DNA sequences encoding distinct alleles.

Composition of the Genome

Typical numbers of genes and size of genomes vary widely among organisms, even those that are fairly closely evolutionarily related. Although it was believed before the completion of the Human Genome Project that the human genome would contain many more genes than simpler animals such as mice or fruit flies, the completion of the project has revealed that the human genome has an unexpectedly low gene density. Estimates of the number of genes in a genome are difficult to compile because they depend on gene finding algorithms that search for patterns resembling those present in known genes, such as open reading frames, promoter regions with sequences resembling the consensus promoter sequence, and related regulatory regions such as TATA boxes in eukaryotes. Gene finding is less reliable in eukaryotic than in prokaryotic genomes due to the presence of non-coding DNA such as introns and pseudogenes. Computational gene finding methods are still significantly more reliable than earlier techniques that required mapping the locations of specific mutations that gave rise to distinguishable alleles.

Gene Content and Genome Size of Various Organisms

Species	*Genome size (Mb)*	*Number of genes*
Mycoplasma genitalium	0.58	500
Streptococcus pneumoniae	2.2	2300
Escherichia coli	4.6	4400
Saccharomyces cerevisiae	12	5800
Arabidopsis thaliana	125	25,500
Caenorhabditis elegans	97	19,000
Sea urchin	814	23,300
Drosophila melanogaster	180	13,700
Mus musculus	2500	29,000
Homo sapiens	2900	27,000
Oryza sativa	466	45-55,000

In most eukaryotic species, very little of the DNA in the genome encodes proteins, and the genes may be separated by vast regions of non-coding DNA, much of which has been labeled "junk DNA" due to its apparent lack of function in the modern organism. A commonly studied type of "junk DNA" is the pseudogenes, or region of non-coding DNA that resembles expressed genes but usually lacks appropriate promoters and other control sequences; such regions are hypothesized to be the results of gene duplication events in a lineage's evolutionary past. Moreover, the genes are often fragmented internally by non-coding sequences called introns, which can be many times longer than the coding sequence but are spliced during post-transcriptional modification of pre-mRNA.

Genetic and Genomic Nomenclature

Gene nomenclature has been established by the HUGO Gene Nomenclature Committee (HGNC) for each known human gene in the form of an approved gene name and symbol (short-form abbreviation). All approved symbols are stored in the HGNC Database. Each symbol is unique and each gene is only given one approved gene symbol. It is necessary to provide a unique symbol for each gene so that people can talk about them. This also facilitates electronic data retrieval from publications. In preference each symbol maintains parallel construction in different members of a gene family and can be used in other species, especially the mouse.

Evolutionary Concept of a Gene

George C. Williams first explicitly advocated the gene-centric view of evolution in his 1966 book *Adaptation and Natural Selection*. He proposed an evolutionary concept of gene to be used when we are talking about natural selection favoring some genes. The definition is: "that which segregates and recombines with appreciable frequency." According to this definition, even an asexual genome could be considered a gene, insofar it have an appreciable permanency through many generations.

The difference is: the molecular gene *transcribes* as a unit, and the evolutionary gene *inherits* as a unit.

Richard Dawkins' *The Selfish Gene* and *The Extended Phenotype* defended the idea that the gene is the only replicator in living systems. This means that only genes transmit their structure largely intact and are potentially immortal in the form of copies. So, genes should be the unit of selection. In *The Selfish Gene* Dawkins attempts to redefine the word 'gene' to mean "an inheritable unit" instead of the generally accepted definition of "a section of DNA coding for a particular protein". In *River Out of Eden*, Dawkins further refined the idea of gene-centric selection by describing life as a river of compatible genes flowing through geological time. Scoop up a bucket of genes from the river of genes, and we have an organism serving as temporary bodies or survival machines. A river of genes may fork into two branches representing two non-interbreeding species as a result of geographical separation.

The *Gene* Concept is Still Changing

The concept of the gene has changed considerably. Originally considered a "unit of inheritance" to a usually DNA-based unit that can exert its effects on the organism through RNA or protein products. It was also previously believed that one gene makes one protein; this concept has been overthrown by the discovery of alternative splicing and trans-splicing.

And the definition of gene is still changing. The first cases of RNA-based inheritance have been discovered in mammals. In plants, cases of traits reappearing after several generations of absence have lead researchers to hypothesise RNA-directed overwriting of genomic DNA. Evidence is also accumulating that the control regions of a gene do not necessarily have to be close to the coding sequence on the linear molecule or even on the same chromosome. Spilianakis and colleagues discovered that the promoter region of the interferon-gamma gene on chromosome 10 and the regulatory regions of the T(H)2 cytokine locus on chromosome 11 come into close proximity in the nucleus possibly to be jointly regulated.

The concept that genes are clearly delimited is also being eroded. There is evidence for fused proteins stemming from two adjacent genes that can produce two separate protein products. While it is not clear whether these fusion proteins are functional, the phenomena is more frequent than previously thought. Even more ground-breaking than the discovery of fused genes is the observation that some proteins can be composed of exons from far away regions and even different chromosomes. This new data has led to an updated, and probably tentative, definition of a gene as "a union of genomic sequences encoding a coherent set of potentially overlapping functional products." This new definition categorizes genes by functional products, whether they be proteins or RNA, rather than specific DNA loci; all regulatory elements of DNA are therefore classified as *gene-associated* regions.

TRANSGENE

A transgene is a gene or genetic material which has been transferred by any of a number of genetic engineering techniques from one organism to another.

In its most precise usage, the term transgene describes a segment of DNA containing a gene sequence which has been isolated from one organism and is introduced into a different organism. This non-native segment of DNA may retain the ability to produce RNA or protein in the transgenic organism or it may alter the normal function of the transgenic organism's genetic code. Typically the DNA is incorporated into the organism's germ line. For example, in higher vertebrates this can be accomplished by injecting the foreign DNA into the nucleus of a fertilized ovum. This technique is routinely used to introduce human disease genes or other genes of interest into strains of laboratory mice to study the function or pathology involved with that particular gene.

In looser usage, transgene can describe any DNA sequence, regardless of whether it contains a gene coding sequence or has been artificially constructed, which has been introduced into an organism or vector construct in which it was previously unknown.

In practical terms, a transgene can be either based on a cDNA (complementary DNA), which is a copy of mRNA (messenger RNA) or on the gene itself residing in its original region of genomic DNA. The difference between these two lies with the facts that the cDNA has been processed to remove introns and also, usually, does not include the regulatory signals that are embedded around and in the gene. The advent of annotated cloned regions of the genome alongside the genome sequence, particularly as large clones in BACs (bacterial artificial chromosomes) or fosmids, and recombineering, which is the method that permits the engineering of these large clones, has changed the practice of transgenesis from its origins with cDNA-based constructs towards the more reliable genomic-based constructs.

PARATRANSGENESIS

Paratransgenesis is a technique that attempts to eliminate a pathogen from vector populations through transgenesis of a symbiont of the vector. The goal of this technique is to control vector-borne diseases. The first step is to identify proteins that prevent the vector species from transmitting the pathogen. The genes coding for these proteins are then introduced into the symbiont, so that they can be expressed in the vector. The final step in the strategy is to introduce these transgenic symbionts into vector populations in the wild.

The first example of this technique used *Rhodnius prolixus* which is associated with the symbiont *Rhodococcus rhodnii*. *Rhodnius prolixus* is an important insect vector of Chagas's disease that is caused by *Trypanosoma cruzi*. The strategy was to engineer *R. rhodnii* to express proteins such as Cecropin A that are toxic to *Trypanosoma cruzi* or that block the transmission of *Trypanosoma cruzi*.

In order to perform paratransgenesis, there are several requirements:

- The Symbiotic bacteria can be grown *in vitro* easily.
- They can be genetically modified, such as through transformation with a plasmid containing the desired gene.
- The engineered symbiont is stable and safe.
- The association between vector and symbiont cannot be attenuated.
- Field delivery is easily handled.

EUGENICS

Eugenics is a social philosophy which advocates the improvement of human hereditary traits through various forms of intervention. Throughout history, eugenics has been regarded by its various advocates as a social responsibility, an altruistic stance of a society, meant to create healthier and more intelligent people, to save resources, and lessen human suffering. More controversially, some, such as the Nazi regime in Germany, used eugenics as a pretext for racial discrimination.

Earlier proposed means of achieving these goals focused on selective breeding, while modern ones focus on prenatal testing and screening, genetic counseling, birth control, in vitro fertilization, and genetic engineering. Opponents argue that eugenics is immoral and is based on, or is itself, pseudoscience.

Historically, eugenics has been used as a justification for coercive state-sponsored discrimination and human rights violations, such as forced sterilization of persons who are claimed to have genetic defects, the killing of the institutionalized population and, in some cases, outright genocide of races perceived as inferior or undesirable.

The modern field and term were first formulated by Sir Francis Galton in 1865, drawing on the recent work of his cousin Charles Darwin. From its inception eugenics was supported by prominent people, including Alexander Graham Bell, George Bernard Shaw, and Margaret Sanger. G. K. Chesterton was against the philosophy of eugenics and he expresses this in his book, *Eugenics and Other Evils*. Eugenics became an academic discipline at many colleges and universities. Funding was provided by prestigious sources such as the Rockefeller Foundation, the Carnegie Institution of Washington, and the Harriman family. Three International Eugenics Conferences presented a global venue for eugenicists with meetings in 1912 in London, and in 1921 and 1932 in New York. Eugenics' scientific reputation started to tumble in the 1930s, a time when Ernst Rüdin began incorporating eugenic rhetoric into the racial policies of Nazi Germany.

Since the postwar period, both the public and the scientific communities have associated eugenics with Nazi abuses, such as enforced racial hygiene, human experimentation, and the extermination of undesired population groups. However, developments in genetic, genomic, and reproductive technologies at the end of the 20th century have raised many new questions and concerns about what exactly constitutes the meaning of *eugenics* and what its ethical and moral status is in the modern era.

Meanings and Types of Eugenics

The word *eugenics* etymologically derives from the Greek word *eu* (*good* or *well*) and the suffix *-gençs* (*born*), and was coined by Sir Francis Galton in 1883.

Eugenics has, from the very beginning, meant many different things to many different people. Historically, the term has referred to everything from prenatal care for mothers to forced sterilization and euthanasia. Much debate took place in the past, and takes place today, as to what exactly counts as eugenics. Some types of eugenics, such as race-based eugenics and class-based eugenics, are sometimes called 'pseudo-eugenics' by proponents of strict eugenics that deals only with perceived beneficial and detrimental genetic traits.

The term *eugenics* is often used to refer to movements and social policies that were influential during the early 20th century. In a historical and broader sense, eugenics can also be a study of "improving human genetic qualities". It is sometimes broadly applied to describe any human action whose goal is to improve the gene pool. Some forms of infanticide in ancient societies, present-day reprogenetics, preemptive abortions and designer babies have been (sometimes controversially) referred to as eugenic.

Because of its normative goals and historical association with scientific racism, as well as the development of the science of genetics, the western scientific community has mostly disassociated itself from the term "eugenics", although one can find advocates of what is now known as *liberal eugenics*.

Ideological social determinists, some of which have obtained college degrees in fields relevant to eugenics, often describe eugenics as a pseudoscience. Modern inquiries into the potential use of genetic engineering have led to an increased invocation of the history of eugenics in discussions of bioethics, most often as a cautionary tale. Some ethicists suggest that even non-coercive eugenics programs would be inherently unethical, though this view has been challenged by such thinkers as Nicholas Agar.

Eugenicists advocate specific policies that (if successful) would lead to a perceived improvement of the human gene pool. Since defining what improvements are desired or beneficial is perceived by many as a cultural choice rather than a matter that can be determined objectively (e.g., by empirical, scientific inquiry), eugenics has often been deemed a pseudoscience. The most disputed aspect of eugenics has been the definition of "improvement" of the human gene pool, such as what is a beneficial characteristic and what is a defect. This aspect of eugenics has historically been tainted with scientific racism.

Early eugenicists were mostly concerned with perceived intelligence factors that often correlated strongly with social class. Many eugenicists took inspiration from the selective breeding of animals (where purebreds are often strived for) as their analogy for improving human society. The mixing of races (or miscegenation) was usually considered as something to be avoided in the name of racial purity. At the time this concept appeared to have some scientific support, and it remained a contentious issue until the advanced development of genetics led to a scientific consensus that the division of the human species into unequal races is unjustifiable. Some see this as an ideological consensus, since equality, just like inequality, is a cultural choice rather than a matter that can be determined objectively.

Eugenics has also been concerned with the elimination of hereditary diseases such as hemophilia and Huntington's disease. However, there are several problems with labeling certain factors as "genetic defects":

- In many cases there is no scientific consensus on what a "genetic defect" is. It is often argued that this is more a matter of social or individual choice.
- What appears to be a "genetic defect" in one context or environment may not be so in another. This can be the case for genes with a heterozygote advantage, such as sickle cell anemia or Tay-Sachs disease, which in their heterozygote form may offer an advantage against, respectively, malaria and tuberculosis.

- Although some birth defects are uniformly lethal, disabled persons can succeed in life.
- Many of the conditions early eugenicists identified as inheritable (pellagra is one such example) are currently considered to be at least partially, if not wholly, attributed to environmental conditions.

Similar concerns have been raised when a prenatal diagnosis of a congenital disorder leads to abortion (see also preimplantation genetic diagnosis).

Eugenic policies have been conceptually divided into two categories:

Positive eugenics is aimed to encourage reproduction among the genetically advantaged. Possible approaches include financial and political stimuli, targeted demographic analyses, in vitro fertilization, egg transplants, and cloning.

Negative eugenics is aimed at lowering fertility among the genetically disadvantaged. This includes abortions, sterilization, and other methods of family planning.

Both positive and negative eugenics can be coercive. Abortion by "fit" women was illegal in Nazi Germany and in the Soviet Union between 1936 and 1968.

During the 20th century, many countries enacted various eugenics policies and programs, including:

- Genetic screening
- Birth control
- Promoting differential birth rates
- Marriage restrictions
- Segregation (both racial segregation as well as segregation of the mentally ill from the normal)
- Compulsory sterilization
- Forced abortions, or, conversely, forced pregnancies
- Genocide

Most of these policies were later regarded as coercive, restrictive, or genocidal, and now few jurisdictions implement policies that are explicitly labeled as eugenic or unequivocally eugenic in substance (however labeled). However, some private organizations assist people in genetic counseling, and reprogenetics may be considered as a form of non-state-enforced "liberal" eugenics.

There are 3 main ways by which the methods of eugenics can be applied. They are:

- mandatory eugenics, in which the government mandates a eugenics program.
- promotional voluntary eugenics, in which eugenics is voluntarily practiced and promoted to the general population, but not officially mandated.
- private eugenics, which is practiced voluntarily by individuals and groups, but not promoted to the general population.

History

Pre-Galtonian Eugenic Philosophies

The basic ideals of eugenics can be found from the beginnings of Western civilization. The philosophy was most famously expounded by Plato, who believed human reproduction should be monitored and controlled by the state. The basic eugenic principle from Plato's The Republic was, "The best men must have intercourse with the best women as frequently as possible, and the opposite is true of the very inferior." Plato understood that this form of government control would not be readily accepted, and proposed that the truth be concealed from the public via a fixed lottery. Mates, in Plato's Republic, would be chosen by a "marriage number" in which the quality of the individual would be quantitatively analyzed, and persons of high numbers would be allowed to procreate with other persons of high numbers. In theory, this would lead to predictable results and the improvement of the human race. However, Plato acknowledged the failure of the "marriage number" since "gold soul" persons could still produce "bronze soul" children. This might have been one of the earliest attempts to mathematically analyze genetic inheritance, which was not perfected until the development of Mendelian genetics and the mapping of the human genome. Other ancient civilizations, such as Rome and Sparta, practiced infanticide as a form of phenotypic selection. In Sparta, newborns were inspected by the city's elders, who decided the fate of the infant. This would be done through abandonment of "weak" or undesirable babies on the slopes of Mount Taygetos, and trials for babies which included bathing them in wine and exposing them to the elements. To Sparta, this would ensure that only the strongest survived and procreated.

The 12 Tables of Roman Law, established early in the formation of the Roman Republic, stated in the fourth table that deformed children would be put to death. In addition, patriarchs in Roman society were given the right to "discard" infants at their discretion. This was often done by drowning undesired newborns in the Tiber River. The practice of infanticide in the ancient world did not subside until the Christianization of the Roman empire.

Galton's Theory

During the 1860s and 1870s, Sir Francis Galton systematized these ideas and practices according to new knowledge about the evolution of man and animals provided by the theory of his cousin Charles Darwin. After reading Darwin's *Origin of Species*, Galton built upon Darwin's ideas whereby the mechanisms of natural selection were potentially thwarted by human civilization. He reasoned that, since many human societies sought to protect the underprivileged and weak, those societies were at odds with the natural selection responsible for extinction of the weakest. Only by changing these social policies, Galton thought, could society be saved from a "reversion towards mediocrity", a phrase that he first coined in statistics and which later changed to the now common "regression towards the mean".

Galton first sketched out his theory in the 1865 article "Hereditary Talent and Character", then elaborated it further in his 1869 book *Hereditary Genius*. He began by studying the way in which human intellectual, moral, and personality traits tended to run in families. Galton's basic argument was that "genius" and "talent" were hereditary traits in humans (although neither he nor Darwin yet had a working model of this type of heredity). He concluded that, since one could use artificial selection to exaggerate traits in other animals, one could expect similar results when applying such models to humans. As he wrote in the introduction to *Hereditary Genius*:

> "I propose to show in this book that a man's natural abilities are derived by inheritance, under exactly the same limitations as are the form and physical features of the whole organic world. Consequently, as it is easy, notwithstanding those limitations, to obtain by careful selection a permanent breed of dogs or horses gifted with peculiar powers of running, or of doing anything else, so it would be quite practicable to produce a highly-gifted race of men by judicious marriages during several consecutive generations."

According to Galton, society already encouraged dysgenic conditions, claiming that the less intelligent were out-reproducing the more intelligent. Galton did not propose any selection methods; rather, he hoped that a solution would be found if social mores changed in a way that encouraged people to see the importance of breeding.

Galton first used the word *eugenic* in his 1883 *Inquiries into Human Faculty and Its Development*, a book in which he meant "to touch on various topics more or less connected with that of the cultivation of race, or, as we might call it, with 'eugenic' questions." He included a footnote to the word "eugenic" which read:

That is, with questions bearing on what is termed in Greek, *eugenes* namely,

> good in stock, hereditarily endowed with noble qualities. This, and the allied words, *eugeneia*, etc., are equally applicable to men, brutes, and plants. We greatly want a brief word to express the science of improving stock, which is by no means confined to questions of judicious mating, but which, especially in the case of man, takes cognisance of all influences that tend in however remote a degree to give to the more suitable races or strains of blood a better chance of prevailing speedily over the less suitable than they otherwise would have had. The word *eugenics* would sufficiently express the idea; it is at least a neater word and a more generalised one than *viriculture* which I once ventured to use.

In 1904 he clarified his definition of eugenics as "the science which deals with all influences that improve the inborn qualities of a race; also with those that develop them to the utmost advantage."

Galton's formulation of eugenics was based on a strong statistical approach, influenced heavily by Adolphe Quetelet's "social physics". Unlike Quetelet, however, Galton did not exalt the "average man" but decried him as mediocre. Galton and his statistical heir Karl Pearson developed what was called the biometrical approach to eugenics, which developed new and complex statistical models (later exported to wholly different fields) to describe the heredity of traits. However, with the rediscovery of Gregor Mendel's hereditary laws, two separate camps of eugenics advocates emerged. One was made up of statisticians, the other of biologists. Statisticians thought the biologists had exceptionally crude mathematical models, while biologists thought the statisticians knew little about biology.

Eugenics eventually referred to human selective reproduction with an intent to create children with desirable traits, generally through the approach of influencing differential birth rates. These policies were mostly divided into two categories: *positive eugenics*, the increased reproduction of those seen to have advantageous hereditary traits; and *negative eugenics*, the discouragement of reproduction by those with hereditary traits perceived as poor. Negative eugenic policies in the past have ranged from attempts at segregation to sterilization and even genocide. Positive eugenic policies have typically taken the form of awards or bonuses for "fit" parents who have another child. Relatively innocuous practices like marriage counseling had early links with eugenic ideology.

Eugenics differed from what would later be known as Social Darwinism. While both claimed intelligence was hereditary, eugenics asserted that new policies were needed to actively change the status quo towards a more "eugenic" state, while the Social Darwinists argued society itself would naturally "check" the problem of "dysgenics" if no welfare policies were in place (for example, the poor might reproduce more but would have higher mortality rates).

Sparta

In Sparta, eugenics was practised, not scientifically but selectively and judged by the infant's ability to live. If the child was deemed incapable of living, it was usually thrown from the Taygetus mountain. It was more common for girls than boys to be killed this way.

Adolf Hitler considered Sparta to be the first "Völkisch State," and much like Ernst Haeckel before him, praised Sparta due to its primitive form of eugenics practice of selective infanticide policy which was applied on deformed children.

Nazi Germany

Nazi Germany under Adolf Hitler was infamous for eugenics programs which attempted to maintain a "pure" German race through a series of programs that ran under the banner of "racial hygiene". Among other activities, the Nazis performed extensive experimentation on live human beings to test their genetic theories, ranging from simple measurement of physical characteristics to the horrific experiments carried out by Josef Mengele for Otmar von Verschuer on twins in the concentration camps. During the 1930s and 1940s, the Nazi regime forcibly sterilized hundreds of thousands of people whom they viewed as mentally and physically "unfit", an estimated 400,000 between 1934 and 1937. The scale of the Nazi program prompted American eugenics advocates to seek an expansion of their program, with one complaining that "the Germans are beating us at our own game". The Nazis went further, however, killing tens of thousands of the institutionalized disabled through compulsory "euthanasia" programs.

They also implemented a number of "positive" eugenics policies, giving awards to "Aryan" women who had large numbers of children and encouraged a service in which "racially pure" single women could deliver illegitimate children. Allegations that such women were also impregnated by SS officers in the *Lebensborn* are common, but unproven. Also, "racially valuable" children from occupied countries were forcibly removed from their parents and adopted by German people. Many of their concerns for eugenics and racial hygiene were also explicitly present in their systematic killing of millions of "undesirable" people including Jews, Gypsies, Jehovah's Witnesses and homosexuals during the Holocaust (much of the killing equipment and methods employed in the death camps were first developed in the euthanasia program). The scope and coercion involved in the German eugenics programs along with a strong use of the rhetoric of eugenics and so-called "racial science" throughout the regime created an indelible cultural association between eugenics and the Third Reich in the postwar years.

The alleged relation between eugenics and the Jewish Holocaust is also questioned. Contrary to popular beliefs Hitler did not regard the Jews as intellectually inferior and did not send them to the concentration camps on these grounds. In fact, in the 1930s Germans regarded the Jews as a highly talented people. Hitler had different reasons for his genocidal policies toward the Jews. Seymour W. Itzkoff writes that the Holocaust was "a vast dysgenic program to rid Europe of highly intelligent challengers to the existing Christian domination by a numerically and politically minuscule minority". Therefore, according to Itzkoff, "the Holocaust was the very antithesis of eugenic practice."

Eugenics and the United States, 1890s–1945

One of the earliest modern advocates of eugenics (before it was labeled as such) was Alexander Graham Bell. In 1881, Bell investigated the rate of deafness on Martha's Vineyard, Massachusetts. From this he concluded that deafness was hereditary in nature and recommended a marriage prohibition against the deaf ("Memoir upon the formation of a deaf variety of the human Race") despite his own marriage to a deaf woman. Like many other early eugenicists, Bell proposed controlling immigration for the purpose of eugenics, and warned that boarding schools for the deaf could possibly be considered as breeding places of a deaf human race.

In 1907, Indiana became the first of more than thirty states to adopt legislation aimed at compulsory sterilization of certain individuals. Although the law was overturned by the Indiana Supreme Court in, the U.S. Supreme Court upheld the constitutionality of a Virginia law allowing for the compulsory sterilization of patients of state mental institutions in 1927.

Beginning with Connecticut in 1896, many states enacted marriage laws with eugenic criteria, prohibiting anyone who was "epileptic, imbecile or feeble-minded" from marrying. In 1898 Charles B. Davenport, a prominent American biologist, began as director of a biological research station based in Cold Spring Harbor where he experimented with evolution in plants and animals. In 1904 Davenport received funds from the Carnegie Institution to found the Station for Experimental Evolution. The Eugenics Record Office opened in 1910 while Davenport and Harry H. Laughlin began to promote eugenics.

During the 20th century, researchers became interested in the idea that mental illness could run in families and conducted a number of studies to document the heritability of such illnesses as schizophrenia, bipolar disorder, and depression. Their findings were used by the eugenics movement as proof for its cause. State laws were written in the late 1800s and early 1900s to prohibit marriage and force sterilization of the mentally ill in order to prevent the "passing on" of mental illness to the next generation. These laws were upheld by the U.S. Supreme Court in 1927 and were not abolished until the mid-20th century. By 1945 over 45,000 mentally ill individuals in the United States had been forcibly sterilized. All in all, 60,000 Americans suffered from sterilization.

In years to come, the ERO collected a mass of family pedigrees and concluded that those who were unfit came from economically and socially poor backgrounds. Eugenicists such as Davenport, the psychologist Henry H. Goddard and the conservationist Madison Grant (all well respected in their time) began to lobby for various solutions to the problem of the "unfit". (Davenport favored immigration restriction and sterilization as primary methods; Goddard favored segregation in his *The Kallikak Family*; Grant favored all of the above and more, even entertaining the idea of extermination.) Though their methodology and research methods are now understood as highly flawed, at the time this was seen as legitimate scientific research. It did, however, have scientific detractors (notably, Thomas Hunt Morgan, one of the few Mendelians to explicitly criticize eugenics), though most of these focused more on what they considered the crude methodology of eugenicists, and the characterization of almost every human characteristic as being hereditary, rather than the idea of eugenics itself.

Some states sterilized "imbeciles" for much of the 20th century. The U.S. Supreme Court ruled in the 1927 *Buck v. Bell* case that the state of Virginia could sterilize those it thought unfit. The most significant era of eugenic sterilization was between 1907 and 1963, when over 64,000 individuals were forcibly sterilized under eugenic legislation in the United States. A favorable report on the results of sterilization in California, the state with the most sterilizations by far, was published in book form by the biologist Paul Popenoe and was widely cited by the Nazi government as evidence that wide-reaching sterilization programs were feasible and humane. When Nazi administrators went on trial for war crimes in Nuremberg after World War II, they justified the mass sterilizations (over 450,000 in less than a decade) by citing the United States as their inspiration.

The idea of "genius" and "talent" is also considered by William Graham Sumner, a founder of the American Sociological Society (now called the American Sociological Association). He maintained that if the government did not meddle with the social policy of *laissez-faire*, a class of genius would rise to the top of the system of social stratification, followed by a class of talent. Most of the rest of society would fit into the class of mediocrity. Those who were considered to be defective (mentally retarded, handicapped, etc.) had a negative effect on social progress by draining off necessary resources. They should be left on their own to sink or swim. But those in the class of delinquent (criminals, deviants, etc.) should be eliminated from society ("Folkways", 1907).

Eugenics is today often associated with racism and both W.E.B. DuBois and

Marcus Garvey supported eugenics or ideas resembling eugenics as a way to reduce African American suffering and improve their stature. However, methods of eugenics were applied to reformulate more restrictive definitions of white racial purity in existing state laws banning interracial marriage: the so-called anti-miscegenation laws. The most famous example of the influence of eugenics and its emphasis on strict racial segregation on such "anti-miscegenation" legislation was Virginia's Racial Integrity Act of 1924. The U.S. Supreme Court overturned this law in 1967 in Loving v. Virginia, and declared anti-miscegenation laws unconstitutional.

With the passage of the Immigration Act of 1924, eugenicists for the first time played a central role in the Congressional debate as expert advisers on the threat of "inferior stock" from eastern and southern Europe. This reduced the number of immigrants from abroad to 15 percent from previous years, to control the number of "unfit" individuals entering the country. The new act, inspired by the eugenic belief in the racial superiority of "old stock" white Americans as members of the "Nordic race" (a form of white supremacy), strengthened the position of existing laws prohibiting race- mixing. Eugenic considerations also lay behind the adoption of incest laws in much of the U.S. and were used to justify many anti-miscegenation laws.

Various authors, notably Stephen Jay Gould, have repeatedly asserted that restrictions on immigration passed in the United States during the 1920s (and overhauled in 1965 with the Immigration and Nationality Act) were motivated by the goals of eugenics, in particular, a desire to exclude races considered to be inferior from the national gene pool. During the early 20th century, the United States and Canada began to receive far higher numbers of Southern and Eastern European immigrants. Influential eugenicists like Lothrop Stoddard and Harry Laughlin (who was appointed as an expert witness for the House Committee on Immigration and Naturalization in 1920) presented arguments that these were inferior races that would pollute the national gene pool if their numbers went unrestricted. It has been argued that this stirred both Canada and the United States into passing laws creating a hierarchy of nationalities, rating them from the most desirable Anglo-Saxon and Nordic peoples to the Chinese and Japanese immigrants, who were almost completely banned from entering the country. However, several people, in particular Franz Samelson, Mark Snyderman and Richard Herrnstein, have argued that, based on their examination of the records of the congressional debates over immigration policy, Congress gave virtually no consideration to these factors. According to these authors, the restrictions were motivated primarily by a desire to maintain the country's cultural integrity against a heavy influx of foreigners. This interpretation is not, however, accepted by most historians of eugenics.

Some who disagree with the idea of eugenics in general contend that eugenics legislation still had benefits. Margaret Sanger (founder of Planned Parenthood of America) found it a useful tool to urge the legalization of contraception. In its time eugenics was seen by many as scientific and progressive, the natural application of knowledge about breeding to the arena of human life. Before the death camps of World War II, the idea that eugenics could lead to genocide was not taken seriously.

Japan

In the early part of the Shôwa era, Japanese governments executed a eugenic policy to limit the birth of children with "inferior" traits, as well as aiming to protect the life and health of mothers.

The *Race Eugenic Protection Law* was submitted from 1934 to 1938 to the Diet. After four amendments, this draft was promulgated as the *National Eugenic Law* in 1940 by the Konoe government. According to the *Eugenic Protection Law* (1948), sterilization could be enforced on criminals "with genetic predisposition to commit crime", patients with genetic diseases such as total color-blindness, hemophilia, albinism and ichthyosis, and mental affections such as schizophrenia, manic-depressiveness and epilepsy.. Mental illnesses were added in 1952.

Canada

In Canada, the eugenics movement took place early in the 20th Century, particularly in Alberta, and was quite popular. The Sexual Sterilization Act of Alberta was enacted in 1928, focusing the movement on the sterilization of mentally deficient individuals, as determined by the Alberta Eugenics Board. The campaign to enforce this action was backed by groups such as the United Farm Women's Group, including key member Emily Murphy.

Individuals were assessed using IQ tests like the Stanford-Binet. This posed a problem to new immigrants arriving in Canada, as many had not mastered the English language, and often their scores denoted them as having impaired intellectual functioning. As a result, many of those sterilized under the Sexual Sterilization Act were immigrants who were unfairly categorized.

The popularity of the eugenics movement peaked during the depression. Individuals sought an explanation for the financial problems of the nation, and the notion of defective breeding became a scapegoat; citizens blamed individuals considered to be subhuman. The end of the Canadian eugenics movement was brought about when the Sexual Sterilization Act was repealed in 1972.

Australia

The policy of removing Aboriginal children from their parents emerged from an opinion based on Eugenics theory in late nineteenth and early twentieth-century Australia that the 'full-blood' tribal Aborigine would be unable to sustain itself, and was doomed to inevitable extinction. An ideology at the time held that mankind could be divided into a civilisational hierarchy. This white supremacist notion supposed that Northern Europeans were superior in civilisation and that Aborigines were inferior. According to this view, the increasing numbers of mixed-descent children in Australia, labelled as 'half-castes' (or alternatively 'crossbreeds', 'quadroons' and 'octoroons'), were widely seen to be a threat to racial purity. In the first half of the twentieth century, this led to policies and legislation that resulted in the removal of children from their parents. The stated aim was to culturally assimilate mixed-descent people into contemporary Australian society. In all states and territories legislation was passed in the early years of the twentieth century which gave Aboriginal protectors guardianship rights over Aborigines up to the age of sixteen or twenty-one. Policemen or other agents of the state (such as Aboriginal Protection Officers), were given the power to locate and transfer babies and children of mixed descent, from their mothers or families or communities into institutions. In these Australian states and territories, half-caste institutions (both government or missionary) were established in the early decades of the twentieth-century for the reception of these separated children. The 2002 movie *Rabbit-Proof Fence* portrays this system and the harrowing consequences of attempting to overcome it.

In 1915 A.O. Neville was appointed the second Western Australia State *Chief Protector of Aborigines*. During the next quarter-century, he presided over the now notorious 'Assimilation' policy of removing mixed-race Aboriginal children from their parents. This policy in turn created the Stolen Generations and set in motion a grieving process that through the now widely accepted concept of trans-generational grief, would affect many generations to come. In 1936 Neville became the *Commissioner for Native Affairs*, a post he held until his retirement in.

Neville believed that biological absorption was the key to 'uplifting the Native race.' Speaking before the Moseley Royal Commission, which investigated the administration of Aboriginals in 1934, he defended the policies of forced settlement, removing children from parents, surveillance, discipline and punishment, arguing that "they have to be protected against themselves whether they like it or not. They cannot remain as they are. The sore spot requires the application of the surgeon's knife for the good of the patient, and probably against the patients will."

In his twilight years Neville continued to actively promote his policy. Towards the end of his career, Neville published *Australia's Coloured Minority*, a text outlining his plan for the biological absorption of aboriginal people into white Australia. It is a classic example of the eugenics policies popular at the time in the Western world.

Sweden

From just prior to World War II until 1975, Sweden forcibly sterilized more than 62,000 people with Herman Lundborg in the lead of the project. Sweden's large-scale eugenics program targeted ethnic or racial minorities, and the mentally ill. As was the case in other programs, ethnicity and race were believed to be connected to mental and physical health. While many Swedes disliked the program, politicians generally supported it; the ruling left supported it more as a means of promoting social health, while amongst the right it was more about racial protectionism. (Not until 1999 did the Swedish government begin paying compensation to the victims and their families.)

Britain

In Britain, eugenics never received significant state funding. Furthermore, its emphasis was more upon class, rather than race. Indeed, Galton expressed these views during a lecture in 1901 in which he placed the British society into groups. These groupings are shown in the figure and indicate the proportion of society falling into each group and their perceived genetic worth. Galton suggested that negative eugenics (i.e. an attempt to prevent them from bearing offspring) should be applied only to those in the lowest social group (the "Undesirables"), while positive eugenics applied to the higher classes. However, he appreciated the worth of the higher working classes to society and industry.

Sterilisation programmes were never legalised, although some were carried out in private upon the mentally ill by clinicians who were in favour of a more widespread eugenics plan. (Sterilisation had, in fact, been carried out to prevent masturbation in mentally ill patients since the 1820s, long before the eugenics movement.) Indeed, those in support of eugenics shifted their lobbying of Parliament from enforced to voluntary sterilization, in the hope of achieving more legal recognition. But leave for the Labour MP Major A. G. Church,to propse a Private Member's Bill in 1931, which would legalise the operation for voluntary sterilization, was rejected by 167 votes to 89.

The popularity of eugenics in Britain was reflected by the fact that only two universities established courses in this field (University College London and Liverpool University), but the position of a professorship in eugenics was never created at either. The Galton Institute, affiliated to UCL, was headed by Galton's protégé, Karl Pearson.

Other Countries

Almost all non-Catholic Western nations adopted some eugenic legislations. In July 1933 Germany passed a law allowing for the involuntary sterilization of "hereditary and incurable drunkards, sexual criminals, lunatics, and those suffering from an incurable disease which would be passed on to their offspring." Two provinces in Canada carried out thousands of compulsory sterilizations, and these lasted into the 1970s. Many First Nations (native Canadians) were targeted, as well as immigrants from Eastern Europe, as the program identified racial and ethnic minorities to be genetically inferior[*citation needed*]. Besides the large-scale program in the United States, other nations included Australia, Norway, France, Finland, Denmark, Estonia, Iceland, and Switzerland with programs to sterilize people the government declared to be mentally deficient. Singapore practiced a limited form of eugenics that involved discouraging marriage between university graduates and the rest through segregation in matchmaking agencies, in the hope that the former would produce better children.

Marginalization after World War II

After the experience of Nazi Germany, many ideas about "racial hygiene" and "unfit" members of society were publicly renounced by politicians and members of the scientific community. The Nuremberg Trials against former Nazi leaders revealed to the world many of the regime's genocidal practices and resulted in formalized policies of medical ethics and the 1950 UNESCO statement on race. Many scientific societies released their own similar "race statements" over the years, and the Universal Declaration of Human Rights, developed in response to abuses during the Second World War, was adopted by the United Nations in 1948 and affirmed, "Men and women of full age, without any limitation due to race, nationality or religion, have the right to marry and to found a family." In continuation, the 1978 UNESCO declaration on race and racial prejudice states that the fundamental equality of all human beings is the ideal toward which ethics and science should converge.

In reaction to Nazi abuses, eugenics became almost universally reviled in many of the nations where it had once been popular (however, some eugenics programs, including sterilization, continued quietly for decades). Many pre-war eugenicists engaged in what they later labeled "crypto-eugenics", purposefully taking their eugenic beliefs "underground" and becoming respected anthropologists, biologists and geneticists in the postwar world (including Robert Yerkes in the U.S. and Otmar von Verschuer in Germany). Californian eugenicist Paul Popenoe founded marriage counseling during the 1950s, a career change which grew from his eugenic interests in promoting "healthy marriages" between "fit" couples.

The American Life League, an opponent of abortion, charges that eugenics was merely "re-packaged" after the war, and promoted anew in the guise of the population-control and environmentalism movements. They claim, for example, that Planned Parenthood was funded and cultivated by the Eugenics Society for these reasons. Julian Huxley, the first Director-General of UNESCO and a founder of the World

Wildlife Fund was also a Eugenics Society president and a strong supporte of eugenics

> [E]ven though it is quite true that any radical eugenic policy will be for many years politically and psychologically impossible, it will be important for UNESCO to see that the eugenic problem is examined with the greatest care, and that the public mind is informed of the issues at stake so that much that now is unthinkable may at least become thinkable. —Julian Huxley

High school and college textbooks from the 1920s through the '40s often had chapters touting the scientific progress to be had from applying eugenic principles to the population. Many early scientific journals devoted to heredity in general were run by eugenicists and featured eugenics articles alongside studies of heredity in nonhuman organisms. After eugenics fell out of scientific favor, most references to eugenics were removed from textbooks and subsequent editions of relevant journals. Even the names of some journals changed to reflect new attitudes. For example, *Eugenics Quarterly* became *Social Biology* in 1969 (the journal still exists today, though it looks little like its predecessor). Notable members of the American Eugenics Society (1922–94) during the second half of the 20th century included Joseph Fletcher, originator of Situational ethics; Dr. Clarence Gamble of the Procter & Gamble fortune; and Garrett Hardin, a population control advocate and author of *The Tragedy of the Commons*.

In the United States, the eugenics movement had largely lost most popular and political support by the end of the 1930s while forced sterilizations mostly ended in the 1960s with the last performed in 1981. Many US states continued to prohibit biracial marriages with "anti-miscegenation laws" such as Virginia's The Racial Integrity Act of 1924, until they were over-ruled by the Supreme Court in 1967 in Loving v. Virginia. The Immigration Restriction Act of 1924, which was designed to limit the immigration of "dysgenic" Italians, and eastern European Jews, was repealed and replaced by the Immigration and Nationality Act in 1965.

However, some prominent academics continued to support eugenics after the war. In 1963 the Ciba Foundation convened a conference in London under the title "Man and His Future," at which three distinguished biologists and Nobel laureates (Hermann Muller, Joshua Lederberg, and Francis Crick) all spoke strongly in favor of eugenics.

A few nations, notably, Canada and Sweden, maintained large-scale eugenics programs, including forced sterilization of mentally handicapped individuals, as well as other practices, until the 1970s.

Modern Eugenics, Genetic Engineering, and Ethical Re-valuation

Beginning in the 1980s, the history and concept of eugenics were widely discussed as knowledge about genetics advanced significantly. Endeavors such as the Human Genome Project made the effective modification of the human species seem possible again (as did Darwin's initial theory of evolution in the 1860s, along with the rediscovery of Mendel's laws in the early 20th century). The difference at the beginning of the 21st century was the guarded attitude towards eugenics, which had become a watchword to be feared rather than embraced.

Suggestions and Ideas

A few scientific researchers such as psychologist Richard Lynn, psychologist Raymond Cattell, and doctor Gregory Stock have openly called for eugenic policies using modern technology, but they represent a minority opinion in current scientific and cultural circles. One attempted implementation of a form of eugenics was a "genius sperm bank" (1980–99) created by Robert Klark Graham, from which nearly 230 children were conceived (the best known donor were Nobel Prize winners William Shockley and J.D.Watson): other Laureates including Maurice Wilkins destroyed their invitation letters.). In the U.S. and Europe, though, these attempts have frequently been criticized as in the same spirit of classist and racist forms of eugenics of the 1930s. Because of its association with compulsory sterilization and the racial ideals of the Nazi Party, the word *eugenics* is rarely used by the advocates of such programs.

Eugenicists have argued that immigration from countries with low national IQ is undesirable. According to Raymond Cattell "when a country is opening its doors to immigration from diverse countries, it is like a farmer who buys his seeds from different sources by the sack, with sacks of different average quality of contents."

Cyprus

A similar screening policy (including prenatal screening and abortion) intended to reduce the incidence of thalassemia exists on both sides of the island of Cyprus. Since the program's implementation in the 1970s, it has reduced the ratio of children born with the hereditary blood disease from 1 out of every 158 births to almost zero.

In the government controlled areas, tests for the gene are compulsory for both partners, prior to marriage.

United States

There are some states that require a blood test prior to marriage. While these tests are typically restricted to the detection of the sexually transmitted disease Syphilis (which was the most common STD at the time these laws were enacted), some partners will voluntarily test for other diseases and genetic incompatibilities.

Harris polls in 1986 and 1992 recorded majority public support for limited forms of germ-line intervention, especially to prevent "children inheriting usually fatal genetic disease".

In 1971, lobbying by the US organisation The International Association for Voluntary Sterilization (AVS), led politicians and officials at the Office for Equal Opportunity to pay for voluntary sterilization of low income Americans for birth-control purposes. AVS also focused on the International community, and its lobbying led to a US foreign policy and funding from the U.S. Agency for International Development to encourage Third World/Developing World countries to utilise abortion and sterilization in order to control their population growth. For further information see EngenderHealth.

Dor Yeshorim

Dor Yeshorim, a program which seeks to reduce the incidence of Tay-Sachs disease, Cystic Fibrosis, Canavan disease, Fanconi anemia, Familial Dysautonomia, Glycogen storage disease, Bloom's Syndrome, Gaucher Disease, Niemann-Pick Disease, and Mucolipidosis IV among certain Jewish communities, is another screening program which has drawn comparisons with liberal eugenics. In Israel, at the expense of the state, the general public is advised to carry out genetic tests to diagnose these diseases before the birth of a baby. If an unborn baby is diagnosed with one of these diseases among which Tay-Sachs is the most commonly known, the pregnancy may be terminated, subject to consent. Most other Ashkenazi Jewish communities also run screening programs because of the higher incidence of genetic diseases. In some Jewish communities, the ancient custom of matchmaking (shidduch) is still practiced, and in order to attempt to prevent the tragedy of infant death which always results from being homozygous for Tay-Sachs, associations such as the strongly observant Dor Yeshorim (which was founded by a rabbi who lost four children to Tay-Sachs with the purpose of preventing others from suffering the same tragedy) test young couples to check whether they carry a risk of passing on fatal conditions. If both the young man and woman are Tay-Sachs carriers, it is common for the match to be broken off. Judaism, like numerous other religions, discourages abortion unless there is a risk to the mother, in which case her needs take precedence. The effort is not aimed at eradicating the hereditary traits, but rather at the occurrence of homozygosity. The actual impact of this program on allele frequencies is unknown, but little impact would be expected because the program does not impose genetic selection. Instead, it encourages disassortative mating.

Ethical Re-Assessment

In modern bioethics literature, the history of eugenics presents many moral and ethical questions. Commentators have suggested the new "eugenics" will come from reproductive technologies that will allow parents to create so-called "designer babies" (what the biologist Lee M. Silver prominently called "reprogenetics"). It has been argued that this "non-coercive" form of biological "improvement" will be predominantly motivated by individual competitiveness and the desire to create "the best opportunities" for children, rather than an urge to improve the species as a whole, which characterized the early 20th-century forms of eugenics. Because of this non-coercive nature, lack of involvement by the state and a difference in goals, some commentators have questioned whether such activities are eugenics or something else altogether. But critics note[*citation needed*] that Francis Galton, did not advocate for coercion when he defined the principles of eugenics. In other words, eugenics does not mean coercion. It is, according to Galton who originated the term, the proper label for bioengineering of "better" human beings.

Daniel Kevles argues that eugenics and the conservation of natural resources are similar propositions. Both can be practiced foolishly so as to abuse individual rights, but both can be practiced wisely.

Some disability activists argue that, although their impairments may cause them pain or discomfort, what really disables them as members of society is a sociocultural system that does not recognize their right to genuinely equal treatment. They express skepticism that any form of eugenics could be to the benefit of the disabled considering their treatment by historical eugenic campaigns.

James D. Watson, the first director of the Human Genome Project, initiated the Ethical, Legal and Social Implications Program (ELSI) which has funded a number of studies into the implications of human genetic engineering (along with a prominent website on the history of eugenics), because:

> In putting ethics so soon into the genome agenda, I was responding to my own personal fear that all too soon critics of the Genome Project would point out that I was a representative of the Cold Spring Harbor Laboratory that once housed the controversial Eugenics Record Office. My not forming a genome ethics program quickly might be falsely used as evidence that I was a closet eugenicist, having as my real long-term purpose the unambiguous identification of genes that lead to social and occupational stratification as well as genes justifying racial discrimination.

Distinguished geneticists including Nobel Prize-winners John Sulston ("I don't think one ought to bring a clearly disabled child into the world") and Watson ("Once you have a way in which you can improve our children, no one can stop it") support genetic screening. Which ideas should be described as "eugenic" are still controversial in both public and scholarly spheres. Some observers such as Philip Kitcher have described the use of genetic screening by parents as making possible a form of "voluntary" eugenics.

Some modern subcultures advocate different forms of eugenics assisted by human cloning and human genetic engineering, sometimes even as part of a new religious movement (see Raëlism, Cosmotheism, or Prometheism). These groups also talk of "neo-eugenics". "conscious evolution", or "genetic freedom".

Behavioral traits often identified as potential targets for modification through human genetic engineering include intelligence, depression, schizophrenia, alcoholism, sexual behavior (and orientation) and criminality.

Criticism

Diseases vs. Traits

While the science of genetics has increasingly provided means by which certain characteristics and conditions can be identified and understood, given the complexity of human genetics, culture, and psychology there is at this point no agreed objective means of determining which traits might be ultimately desirable or undesirable. Eugenic manipulations that reduce the propensity for criminality and violence, for example, might result in the population being enslaved by an outside aggressor it can no longer defend itself against. On the other hand, genetic diseases like hemochromatosis can increase susceptibility to illness, cause physical deformities, and other dysfunctions. Eugenic measures against many of these diseases are already being undertaken in societies around the world, while measures against traits that affect more subtle, poorly understood traits, such as criminality, are relegated to the realm of speculation and science fiction. The effects of diseases are essentially wholly negative, and societies everywhere seek to reduce their impact by various means, some of which are eugenic in all but name. The other traits that are discussed have positive as well as negative effects and are not generally targeted at present anywhere.

Slippery Slope

A common criticism of eugenics is that it inevitably leads to measures that are unethical (Lynn 2001). A hypothetical scenario posits that if one racial minority group is on average less intelligent than the racial majority group, then it is more likely that the racial minority group will be submitted to a eugenics program rather than the least intelligent members of the whole population.

H.L. Kaye wrote of "the obvious truth that eugenics has been discredited by

Hitler's crimes," (Kaye 1989). R. L. Hayman argued "the eugenics movement is an anachronism, its political implications exposed by the Holocaust," (Hayman 1990).

Steven Pinker has stated that it is "a conventional wisdom among left-leaning academics that genes imply genocide." He has responded to this "conventional wisdom" by comparing the history of Marxism, which had the opposite position on genes to that of Nazism:

> But the 20th century suffered "two" ideologies that led to genocides. The other one, Marxism, had no use for race, didn't believe in genes and denied that human nature was a meaningful concept. Clearly, it's not an emphasis on genes or evolution that is dangerous. It's the desire to remake humanity by coercive means (eugenics or social engineering) and the belief that humanity advances through a struggle in which superior groups (race or classes) triumph over inferior ones.

Richard Lynn broadens his criticism of eugenics, by arguing that any social philosophy is capable of ethical misuse. Though Christian principles have aided in the abolition of slavery and the establishment of welfare programs, he notes that the Christian church has also burned many dissidents at the stake and allowed for the killing of large numbers of innocent people by Crusaders. Lynn argues the appropriate response is to condemn these killings, but believes Christianity does not "inevitably [lead] to the extermination of those who do not accept its doctrines," (Lynn 2001).

Genetic Diversity

Eugenic policies could also lead to loss of genetic diversity, in which case a culturally accepted improvement of the gene pool would very likely, as evidenced in numerous instances in isolated island populations (e.g. the Dodo, Raphus cucullatus, of Mauritius) result in extinction due to increased vulnerability to disease, reduced ability to adapt to environmental change and other factors both known and unknown. A long-term eugenics plan might lead to a scenario similar to this because the elimination of traits deemed undesirable would reduce genetic diversity by definition. (Galton 2001, 48).

Proponents of eugenics argue that in any one generation any realistic program would make only minor changes in the gene pool, giving plenty of time to reverse direction if unintended consequences emerge, reducing the likelihood of the elimination of desirable genes. Proponents of eugenics argue that any appreciable reduction in diversity is so far in the future that little concern is needed for now.

The possible elimination of the autism genotype is a significant political issue in the autism rights movement, which claims autism is a form of neurodiversity. Many advocates of Down Syndrome rights also consider Down Syndrome (Trisomy-21) a form of neurodiversity.

Heterozygous Recessive Traits

In some instances efforts to eradicate certain single-gene mutations would be nearly impossible. In the event the condition in question was a heterozygous recessive trait, the problem is that by eliminating the visible unwanted trait, there are still as many genes for the condition left in the gene pool as were eliminated according to the Hardy-Weinberg principle, which states that a population's genetics are defined as pp+2pq+qq at equilibrium. With genetic testing it may be possible to detect all of the heterozygous recessive traits, but only at great cost with the current technology. Under normal circumstances it is only possible to eliminate a dominant allele from the gene pool. Recessive traits can be severely reduced, but never eliminated unless the complete genetic makeup of all members of the pool was known, as aforementioned. As only very few undesirable traits, such as Huntington's disease, are dominant, the practical value for "eliminating" traits is quite low.

However, there are examples of eugenic acts that managed to lower the prevalence of recessive diseases, although not influencing the prevalence of heterozygote carriers of those diseases. The elevated prevalence of certain genetically transmitted diseases among the Ashkenazi Jewish population (Tay-Sachs, Cystic Fibrosis, Canavan's disease and Goucher's disease), has been decreased in current populations by the application of genetic screening.

Counterarguments

Dysgenics

Some supporters of eugenics allege that a dysgenic decline in intelligence is occurring, which may lead to the collapse of civilization, and justify eugenic programs on that basis.

Potential Benefits

Small differences in average IQ at the group level might theoretically have large effects on social outcomes. Richard Herrnstein and Charles Murray altered the mean IQ (100) of the U.S. National Longitudinal Survey of Youth's population sample by randomly deleting individuals below an IQ of 103 until the population mean reached 103. This calculation was conducted twice and averaged together to avoid error from the random selection. This test showed that the new group with an average IQ of 103 had a poverty rate 25% lower than a group with an average IQ of 100. Similar substantial correlations in high school drop-out rates, crime rates, and other outcomes were measured.

Indeed, many studies suggest that IQ correlates with various socioeconomic factors. However, to what extent IQ is a cause of these socioeconomic factors, as opposed to a consequence of them, is disputed. Studies have suggested, for example, that education increases an individual's IQ — although other studies have shown that education has little to no effect.

LIST OF GENETICS RESEARCH ORGANIZATIONS

This is a list of organizations involved in genetics research or related activities.

Companies

- Affymetrix, UK
- Applera, Norwalk, CT
- Applied Biosystems, Foster City, CA
- Asper Biotech
- Celera Genomics
- Genentech, San Francisco, CA
- Genetix, Hampshire, UK

Research Institutes

Americas

- Arizona, USA
  - Translational Genomics Research Institute (TGen), Phoenix founded in July of 2002
- Arkansas, USA
  - Whitetail Genetic Research Institute, Jerusalem,
- California, USA
  - Genetic Information Research Institute, Mountain View founded 1994
  - Stanford University, Stanford, California
- Florida, USA
  - University of Florida Genetics Institute, Miami
- Maryland, USA
  - The Institute for Genomic Research (TIGR), founded in 1992 by Craig Venter

  - Howard Hughes Medical Institute, Chevy Chase
- Massachusetts, USA
  - Whitehead Institute Center for Genome Research, Cambridge,
- Missouri, USA
  - Genome Sequencing Center, St Louis,
- New York, USA
  - Cold Spring Harbor Laboratory, Cold Spring Harbor, NY,
  - Center for Functional Genomics, SUNY Albany, Rensselaer, NY,
- Washington, DC, USA
  - Research Institute for Genetic and Human Therapy - Instituto di Ricerca per la Terapia Genetica Umana,
- USA
  - Center for the Advancement of Genomics
  - National Human Genome Research Institute founded 1989

Europe

- Austria
  - Institut für Forstgenetik, Vienna
- Germany
  - Zentrum für genetische Forschung, (part of Max-Planck-Institute für Psychiatrie und Biochemie) München
  - Institut für Humangenetik, Hamburg
  - Institut für Humangenetik, (Klinikum der Ruprecht-Karls-Universität Heidelberg), Heidelberg
  - Institut für Humangenetik, (Friedrich-Alexander-Universität), Erlangen-Nürnberg
  - Institut für Humangenetik (Friedrich-Wilhelms-Universität), Bonn
  - Institut für Humangenetik, (University of Lübeck), Lübeck
  - Max Planck Institute for Molecular Genetics, Berlin
  - European Molecular Biology Lab (EMBL), Heidelberg
- Italy
  - International Plant Genetic Resources Institute, Maccarese (Fiumicino)

Global Crop Diversity Trust www.croptrust.org

- UK
  - Medical Research Council Human Genetics Unit, Edinburgh, Scotland
  - Department of Genetics, University of Cambridge
  - Imperial College Genetics and Genomics Research Institute, Hammersmith Hospital, London

Asia/Pacific

- Australia
  - Commonwealth Scientific and Industrial Research Organisation
- India
  - Centre for population genetics, KolKatta
  - Department of Genetics, Dr. ALM. PG. IBMS, University of Madras
- Hong Kong
  - Chinese University of Hong Kong,
  - Hong Kong Institute of Biotechnology,
  - Institute of Molecular Biology, Hong Kong,
- China
  - Beijing Genomics Institute, Beijing, Beijing Genomics Institute website
  - The National Human Genome Centre in Southern China, Shanghai
  - The National Human Genome Centre in Northern China, Beijing
  - Huada Human Genome Research Centre
  - Bio Island Haizhu, Guangzhou
  - The Human Genome Centre of the Chinese Academy of Science (CAS)
- Singapore
  - Biopolis, Singapore
  - Genome Institute of Singapore
  - Bioinformatik Institute, Singapore
  - Institute of Molecular and Cell Biology
  - Bioinformatics Center, Singapore
- Taiwan
  - National Health Research Institutes

Africa

- Kenya
  - International Livestock Research Institute (ILRI), Nairobi,

Genetic Research Watchdog Organizations

- GeneWatch, UK
- Council for Responsible Genetics, US
- Smallpox Bio Security Conference 21-22 Oct 2003, Geneva, Switzerland
- Sunshine Project, Hamburg, Germany and Austin, Texas

REFERENCES

Conner, J.K. and Hartl, D.L. "A Primer of Ecological Genetics". Sinauer Associates, Inc.; Sunderland, Mass. (2004) Provides basic and intermediate level processes and methods.

J. Beatty. "The synthesis and the synthetic theory" in Integrating Scientific Disciplines, edited by W. Bechtel and Nijhoff. Dordrecht, 1986.

Luigi Luca Cavalli-Sforza. Genes, Peoples, and Languages. North Point Press, 2000.

Luigi Luca Cavalli-Sforza et al. The History and Geography of Human Genes. Princeton University Press, 1994.

James F. Crow and Motoo Kimura. Introduction to Population Genetics Theory. Harper & Row, 1972.

Warren J Ewens. Mathematical Population Genetics. Springer-Verlag New York, Inc., 2004.

Spencer Wells. The Journey of Man. Random House, 2002.

Spencer Wells. Deep Ancestry: Inside the Genographic Project. National Geographic Society, 2006.

4

Genome & Genomics: Major Areas of Study & Project

GENOMICS

Genomics is the study of an organism's entire genome. The field includes intensive efforts to determine the entire DNA sequence of organisms and fine-scale genetic mapping efforts. The field also includes studies of intragenomic phenomena such as heterosis, epistasis, pleiotropy and other interactions between loci and alleles within the genome. In contrast, the investigation of single genes, their functions and roles, something very common in today's medical and biological research, and a primary focus of molecular biology, does not fall into the definition of genomics, unless the aim of this genetic, pathway, and functional information analysis is to elucidate its effect on, place in, and response to the entire genome's networks.

History of the Field

Genomics can be said to have appeared in the 1980s, and took off in the 1990s with the initiation of genome projects for several biological species. A major branch of genomics is still concerned with sequencing the genomes of various organisms, but the knowledge of full genomes has created the possibility for the field of functional genomics, mainly concerned with patterns of gene expression during various conditions. The most important tools here are microarrays and bioinformatics. Study of the full set of proteins in a cell type or tissue, and the changes during various conditions, is called proteomics.

In 1972, Walter Fiers and his team at the Laboratory of Molecular Biology of the University of Ghent (Ghent, Belgium) were the first to determine the sequence of a gene: the gene for Bacteriophage MS2 coat protein. In 1976, the team determined the complete nucleotide-sequence of bacteriophage MS2-RNA. The first DNA-based genome to be sequenced in its entirety was that of bacteriophage Ö-X174; (5,368 bp), sequenced by Frederick Sanger in 1977. The first free-living organism to be sequenced was that of *Haemophilus influenzae* (1.8 Mb) in 1995, and since then genomes are being sequenced at a rapid pace. A rough draft of the human genome was completed by the Human Genome Project in early 2001, creating much fanfare.

As of September 2007, the complete sequence was known of about 1879 viruses, 577 bacterial species and roughly 23 eukaryote organisms, of which about half are fungi. Most of the bacteria whose genomes have been completely sequenced are problematic disease-causing agents, such as *Haemophilus influenzae*. Of the other sequenced species, most were chosen because they were well-studied model organisms or promised to become good models. Yeast (*Saccharomyces cerevisiae*) has long been an important model organism for the eukaryotic cell, while the fruit fly *Drosophila melanogaster* has been a very important tool (notably in early pre-molecular genetics). The worm *Caenorhabditis elegans* is an often used simple model for multicellular organisms. The zebrafish *Brachydanio rerio* is used for many developmental studies on the molecular level and the flower *Arabidopsis thaliana* is a model organism for flowering plants. The Japanese pufferfish (*Takifugu rubripes*) and the spotted green pufferfish (*Tetraodon nigroviridis*) are interesting because of their small and compact genomes, containing very little non-coding DNA compared to most species. The mammals dog (*Canis familiaris*), brown rat (*Rattus norvegicus*), mouse (*Mus musculus*), and chimpanzee (*Pan troglodytes*) are all important model animals in medical research.

Bacteriophage Genomics

Bacteriophages have played and continue to play a key role in bacterial genetics and molecular biology. Historically, they were used to define gene structure and gene regulation. Also the first genome to be sequenced was a bacteriophage. However, bacteriophage research did not lead the genomics revolution, which is clearly dominated by bacterial genomics. Only very recently has the study of bacteriophage genomes become prominent, thereby enabling researchers to understand the mechanisms underlying phage evolution. Bacteriophage genome sequences can be obtained through direct sequencing of isolated bacteriophages, but can also be derived as part of microbial genomes. Analysis of bacterial genomes has shown that a substantial amount of microbial DNA consists of prophage sequences and prophage-like elements. A detailed database mining of these sequences offers insights into the role of prophages in shaping the bacterial genome.

Cyanobacteria Genomics

At present there are 24 cyanobacteria for which a total genome sequence is available. 15 of these cyanobacteria come from the marine environment. These are six *Prochlorococcus* strains, seven marine *Synechococcus* strains, *Trichodesmium erythraeum* IMS101 and *Crocosphaera watsonii* [[WH8501. Several studies have demonstrated how these sequences could be used very successfully to infer important ecological and physiological characteristics of marine cyanobacteria. However, there are many more genome projects currently in progress, amongst those there are further *Prochlorococcus* and marine *Synechococcus* isolates, *Acaryochloris* and *Prochloron*, the N_2-fixing filamentous cyanobacteria *Nodularia spumigena*, *Lyngbya aestuarii* and *Lyngbya majuscula*, as well as bacteriophages infecting marine cyanobaceria. Thus, the growing body of genome information can also be tapped in a more general way to address global problems by applying a comparative approach. Some new and exciting examples of progress in this field are the identification of genes for regulatory RNAs, insights into the evolutionary origin of photosynthesis, or estimation of the contribution of horizontal gene transfer to the genomes that have been analyzed.

COMPUTATIONAL GENOMICS

Computational genomics is the study of deciphering biology from genome sequences using computational analysis., including both DNA and RNA. Computational genomics focuses on understanding the human genome, and more generally the principles of how DNA controls the biology of any species at the molecular level. With the current abundance of massive biological datasets, computational studies have become one of the most important means to biological discovery.

History

Computational genomics began in spirit, if not in name, during the 1960s with the research of Margaret Dayhoff and others at the National Biomedical Research Foundation, who first assembled a database of protein sequences. Their research developed a phylogenetic tree that determined the evolutionary changes that were required for a particular protein to change into another protein based on the underlying amino acid sequences. This led them to create a scoring matrix that assessed the likelihood of one protein being related to another.

Beginning in the 1980s, databases of genome sequences began to be recorded, but this presented new challenges in the form of searching and comparing the databases of gene information. Unlike text-searching algorithms that are used on websites such as google or Wikipedia, searching for sections of genetic similarity requires one to find strings that are not simply identical, but similar. This led to the development of the Needleman-Wunsch algorithm, which is a dynamic programming algorithm for comparing sets of amino acid sequences with each other by using scoring matrices derived from the earlier research by Dayhoff. Later, the BLAST algorithm was developed for performing fast, optimized searches of gene sequence databases. BLAST and its derivatives are probably the most widely-used algorithms for this purpose.

The first meeting of the Annual Conference on Computational Genomics was in 1998, providing a forum for this speciality and effectively distinguishing this area of science from the more general fields of Genomics or Computational Biology. The first use of this term in scientific literature, according to MEDLINE abstracts, was just one year earlier in Nucleic Acids Research.

The development of computer-assisted mathematics (using products such as Mathematica or Matlab) has helped engineers, mathematicians and computer scientists to start operating in this domain, and a public collection of case studies and demonstrations is growing, ranging from whole genome comparisons to gene expression analysis. This has increased the introduction of different ideas, including concepts from systems and control, information theory, strings analysis and data mining. It is anticipated that computational approaches will become and remain a standard topic for research and teaching, while students fluent in both topics start being formed in the multiple courses created in the past few years.

Contributions of Computational Genomics Research to Biology

Contributions of computational genomics research to biology include:

- discovering subtle patterns in genomic sequences
- proposing cellular signalling networks
- proposing mechanisms of genome evolution
- predict precise locations of all human genes using [comparative genomics] techniques with several mammalian and vertebrate species
- predict conserved genomic regions that are related to early embryonic development
- discover potential links between repeated sequence motifs and tissue-specific gene expression
- measure regions of genomes that have undergone unusually rapid evolution

STRUCTURAL GENOMICS

Structural genomics consists in the determination of the three dimensional structure of all proteins of a given organism, by experimental methods such as X-ray crystallography, NMR spectroscopy or computational approaches such as homology modelling.

As opposed to traditional structural biology, the determination of a protein structure through a structural genomics effort often (but not always) comes before anything is known regarding the protein function. This raises new challenges in structural bioinformatics, i.e. determining protein function from its 3D structure.

One of the important aspects of structural genomics is the emphasis on high throughput determination of protein structures. This is performed in dedicated centers of structural genomics.

While most structural biologists pursue structures of individual proteins or protein groups, specialists in structural genomics pursue structures of proteins on a genome wide scale. This implies large scale cloning, expression and purification. One main advantage of this approach is economy of scale. On the other hand, the scientific value of some resultant structures is at times questioned. An interesting Science article from January 2006 analyzes the structural genomics field.

The main advantage of the wide spread of structural genomics ceters is that the scientific community rapidly gets access to very many novel protein structures. A small disadvantage is that the great speed of solving these structures precludes extensive manual inspection, which consequently can lead to the occurrence of occasional errors in the structures. A database of all errors (no matter how small or insignificant) detected in PDB files has been produced using the freely available WHAT_CHECK software.

METAGENOMICS

Metagenomics (also Environmental Genomics, Ecogenomics or Community Genomics) is the study of genetic material recovered directly from environmental samples. Traditional microbiology and microbial genome sequencing rely upon cultivated clonal cultures. This relatively new field of genetic research enables studies of organisms that are not easily cultured in a laboratory as well as studies of organisms in their natural environment.

Early environmental gene sequencing cloned specific genes (often the 16s rRNA gene) to produce a profile of diversity in a natural sample. Such work revealed that the vast majority of microbial diversity had been missed by cultivation-based methods. Recent studies use "shotgun" Sanger sequencing or chip-based pyrosequencing to get (mostly) unbiased samples of all genes from all members of sampled communities.

History

Origin of the Term

The term "metagenomics" was first used by Jo Handelsman and others in the University of Wisconsin Department of Plant Pathology, and first appeared in publication in 1998. The term metagenome referenced the idea that a collection of genes sequenced from the environment could be analyzed in way analogous to the study of a single genome. The exploding interest in environmental genetics, along with the buzzword-like nature of the term, has resulted in the broader use of metagenomics to describe any sequencing of genetic material from environmental (i.e. uncultured) samples, even work that focuses on one organism or gene. Recently, Kevin Chen and Lior Pachter (researchers at the University of California, Berkeley) defined metagenomics as "the application of modern genomics techniques to the study of communities of microbial organisms directly in their natural environments, by passing the need for isolation and lab cultivation of individual species."

Environmental Gene Surveys

Conventional sequencing begins with a culture of identical cells as a source of DNA. However early metagenomic studies revealed that there are probably large groups of microorganisms in many environments that cannot be cultured and thus cannot be sequenced. These early studies focused on 16S ribosomal RNA sequences which are relatively short, often conserved within a species, and generally different between species. Many 16S rRNA sequences have been found which do not belong to any known cultured species, indicating that there are numerous unisolated organisms out there.

Early molecular work in the field was conducted by Norman R. Pace and colleagues, who used PCR to explore the diversity of ribosomal RNA sequences. The insights gained from these breakthrough studies led Pace to propose the idea of cloning DNA directly from environmental samples as early as 1985. This led to the first report of isolating and cloning bulk DNA from an environmental sample, published by Pace and colleagues in 1991 while Pace was in the Department of Biology at Indiana University. Considerable efforts ensured that these were not PCR false positives and supported the existence of a complex community of unexplored species. Although this methodology was limited to exploring highly conserved, non-protein coding genes, it did support early microbial morphology-based observations that diversity was far more complex than was known by culturing methods.

Soon after that, Healy reported the metagenomic isolation of functional genes from "zoolibraries" constructed from a complex culture of environmental organisms grown in the laboratory on dried grasses in 1995. After leaving the Pace laboratory, Ed DeLong continued in the field and has published work that has largely laid the groundwork for environmental phylogenies based on signature 16S sequences, beginning with his group's construction of libraries from marine samples.

Longer Sequences From Environmental Samples

Recovery of DNA sequences longer than a few thousand base pairs from environmental samples was very difficult until recent advances in molecular biological techniques, particularly related to constructing libraries in bacterial artificial chromosomes (BACs), provided better vectors for molecular cloning.

Shotgun Metagenomics

Advances in bioinformatics, refinements of DNA amplification, and proliferation of computational power have greatly aided the analysis of DNA sequences recovered from environmental samples. These advances have enabled the adaptation of shotgun sequencing to metagenomic samples. The approach, used to sequence many cultured microorganisms as well as the human genome, randomly shears DNA, sequences many short sequences, and reconstructs them into a consensus sequence. In 2002, Mya Breitbart, Forest Rohwer, and colleagues used environmental shotgun sequencing to show that 200 liters of seawater contains over 5000 different viruses. Subsequent studies showed that there are >1000 viral species in human stool and possibly a million different viruses per kilogram of marine sediment, including many bacteriophages. Essentially all of the viruses in these studies were new species. A 2004 metagenomic study of the Sargasso Sea found DNA from nearly 2000 different species including 148 types of bacteria never seen before.

Also in 2004, Gene Tyson, Jill Banfield, and colleagues at the University of California, Berkeley and the Joint Genome Institute sequenced DNA extracted from an acid mine drainage system. This effort resulted in the complete, or nearly complete, genomes for a handful of bacteria and archaea that had previously resisted attempts to culture them. It was now possible to study entire genomes without the biases associated with laboratory cultures.

In 2006 Robert Edwards, Forest Rohwer, and colleagues at San Diego State University published the first sequences of environmental samples generated with so-called next generation sequencing, in this case chip based pyrosequencing developed by 454 Life Sciences. This technique for sequencing the DNA generates shorter fragments than the conventional techniques, however this limitation is compensated for by the very large number of sequences generated. In addition, this technique does not require cloning the DNA before sequencing, removing one of the main biases in metagenomics.

Microbial Diversity

Much of the interest in metagenomics comes from the discovery that the vast majority of microorganisms had previously gone unnoticed. Traditional microbiological methods relied upon laboratory cultures of organisms. Surveys of ribosomal RNA (rRNA) genes taken directly from the environment revealed that cultivation based methods find less than 1% of the bacteria and archaea species in a sample.

Gene Surveys

Shotgun sequencing and screens of clone libraries reveal genes present in environmental samples. This provides information both on which organisms are present and what metabolic processes are possible in the community. This can be helpful in understanding the ecology of a community, particularly if multiple samples are compared to each other.

Environmental Genomes

Shotgun metagenomics also is capable of sequencing nearly complete microbial genomes directly from the environment. Because the collection of DNA from an environment is largely uncontrolled, the most abundant organisms types in a sample are most highly represented in the resulting sequence data. To achieve the high coverage needed to fully resolve the genomes of underrepresented community members, large samples, often prohibitively so, are needed. On the other hand, the random nature of shotgun sequencing ensures that many of these organisms will be represented by at least some small sequence segments. Due to the limitations of microbial isolation methods, the vast majority of these organisms would go unnoticed using traditional culturing techniques.

Community Metabolism

Many bacterial communities show significant division of labor in metabolism. Waste products of some organisms are metabolites for others. Working together they turn raw resources into fully metabolized waste. Using comparative gene studies and expression experiments with microarrays or proteomics researchers can piece together a metabolic network that goes beyond species boundaries. Such studies require detailed knowledge about which versions of which proteins are coded by which species and even by which strains of which species. Therefore, community genomic information is another fundamental part (as metabolomics or proteomics) to be able to estimate how metabolites are possibly transferred and transformed through a community.

PERSONAL GENOMICS

Personal genomics is a genomics field to popularize the technology to common people. The main goal of personal genomics is to sequence as many people as possible to do a very large scale comparative genomics analysis. Once a great number of 3 billion bases of individuals are collected, human genetic diversity will be completely mapped. This will lead to personal medicine. There have been several major advocates of this genomics subfield. George Church group in Harvard Medical School has been active in this area in the USA.

TOXICOGENOMICS

Toxicogenomics is a form of analysis by which the activity of a particular toxin or chemical substance on living tissue can be identified based upon a profiling of its known effects on genetic material. Once viable, the technique should serve for toxicology and toxin-determination a role analogous to DNA-testing in the forensic identification of individuals.

Toxicogenomics may also be of use as a preventative measure to predict adverse "side", i.e. toxic, effects, of pharmaceutical drugs on susceptible individuals. This involves using genomic techniques such as gene expression level profiling and single-nucleotide polymorphism analysis of the genetic variation of individuals.

Studies of those types are then correlated to adverse toxicological effects in clinical trials so that suitable diagnostic markers (measurable signs) for these adverse effects can be developed.

Using such methods, it would then be theoretically possible to test an individual patient for his or her susceptibility to these adverse effects before administering a drug. Patients that would show the marker for an adverse effect would be switched to a different drug. While this approach is currently theoretical, it has great potential.

There are many well-publicized cases in which popular drugs such as Vioxx and fen-phen were pulled from the market because of toxic effects experienced by a small percentage of patients, with a cost of many billions of dollars to the companies responsible, and the loss of a helpful drug to individuals not at risk for the side effects. If an accurate test using toxicogenomic methods could be developed that successfully identified patients who are susceptible to these adverse effects, these drugs could be returned to market with very little risk.

Such would have the triple benefit of re-allowing the therapeutic use of a previously banned drug, preventing potentially life-threatening side effects, and restoring the majority of the lost market share of these drugs to the company that developed them.

PALEOPOLYPLOIDY

Paleopolyploidy refers to ancient genome duplications which occurred at least several million years ago (mya). The genome doubling event could either be an autopolyploidy or an allopolyploidy. Due to functional redundancy, genes are rapidly silenced and/ or lost from the duplicated genomes. Most paleopolyploids, through evolutionary time, have lost their "polyploid" status through a process called diploidization, and are referred as "diploids" nowadays (eg. baker's yeast, *Arabidopsis* and perhaps humans).

Paleopolyploidy is extensively studied in plant linages. It has been found that almost all flowering plants have at least one round of genome duplication sometime during their evolutionary history. Ancient genome duplications are also found in the early ancestor of vertebrates (which includes the human lineage) and another near the origin of the bony fishes. Interestingly, evidence suggests that baker's yeast (*Saccharomyces cerevisiae*), which has a compact genome, experienced polyploidzation during its evolutionary history.

Paleopolyploidy in Eukaryotes

Ancient genome duplications are widespread throughout eukaryotic lineages, particularly in plants. Almost all important cereal crops are paleopolyploids. Studies suggest that the common ancestor of Poaceae, the grass family, had a genome duplication 50-70 mya. Subsequent genome doublings occurred in maize and wheat (two rounds).

Furthermore, *Arabidopsis thaliana*, which has a small genome for a plant, experienced at least two rounds of paleopolyploidy. The most recent event took place before the divergence of *Arabidopsis* and *Brassica* lineages 25-40 mya; another one that is shared by all eudicots 50-70 mya and perhaps a third round that is shared by all flowering plants >200 mya.

Compared with plants, paleopolyploidy is much rarer in the animal kingdom. It is identified mainly in the amphibians and bony fishes. Although some studies suggested one (sometimes two) common genome duplications are shared by all vertebrates (including human), the evidence is not as strong as other cases and it is still under debate. However, many researchers were interested in the reason why animal lineages had fewer paleopolyploidization events than plants.

Lastly, a well-supported paleopolyploidy has been found in baker's yeast (*Saccharomyces cerevisiae*), despite its small, compact genome (~13Mbp) after the divergence from *K. waltii*. Through genome streamlining, yeast has lost 90% of the duplicated genome over evolutionary time and is recognized as a diploid organism nowadays.

Detection Method

Duplicated genes can be identified through sequence homology on the DNA or protein level. Paleopolyploidy can be identified as massive gene duplication at one time using a molecular clock. To distinguish between whole-genome duplication and a collection of single gene duplication (which is a common phenomenon in the genome) events, the following rules are often applied:

1. Duplicated genes are located in large duplicated blocks. Single gene duplication is a random process and tends to make duplicated genes scattered throughout the genome.
2. Duplicated blocks are non-overlapping because they were created simultaneously Segmental duplication within the genome can fulfill Rule #1; but multiple independent segmental duplications could overlap each other.

In theory, the two duplicated genes should have the same "age"; that is, the divergence of the sequence should be equal between the two genes duplicated by paleopolyploidy (homeologs). Synonymous substitution rate, Ks, is often used as a molecular clock to determine the time of gene duplication. Thus, paleopolyploidy is identified as a "peak" on the duplicate number vs. Ks graph (shown on the right).

Duplication events that occurred a long time ago in the history of various evolutionary lineages can be difficult to detect because of subsequent diploidization (such that a polyploid starts to behave cytogenetically as a diploid over time) as mutations and gene translations gradually make one copy of each chromosome unlike its counterpart. This usually results in a low confidence for identifying a very ancient paleopolyploidy.

Evolutionary Importance

Paleopolyploidization events lead to massive cellular changes, including doubling of the genetic material, changes in gene expression and increased cell size. Genes lost during diploidization is not completely random, but heavily selected. Genes from large gene family, which diversity is required for their functions or genes that function in a dosage manner, tend to be retained (eg. cellular machinery, transcription factors and protein kinases). On the other hand, genes involved in DNA repair, apoptosis and transmembrane receptors tend to return to a single copy status. Overall, paleopolyploidy can have both short-term and long-term evolutionary effects on an organism's fitness in the natural environment.

- Genome Diversity — genome doubling provided the organism with redundant alleles that can evolve freely with little selection pressure. The duplicated genes can undergo neofunctionalization or subfunctionalization which could help the organism adapt to the new environment or survive different stress conditions.

- Heterosis — polyploids often have larger cell sizes and even larger organs. Many important crops, including wheat, maize and cotton, are paleopolyploids which were domesticated by ancient people.
- Speciation — It has been suggested that many polyploidization events created new species because of gaining adaptive traits or sexual incompatibility with their diploid counterparts. An example would be the recent speciation of allopolyploid Spartina — S. anglica; the polyploid plant is so successful that it is listed as invasive species in many regions.

Human as Paleopolyploid

The hypothesis about human being paleopolyploid originated as early as 1970, proposed by the famous biologist Susumu Ohno. He reasoned that vertebrate genome can not achieve its complexity without large scale whole-genome duplications. The "two rounds of genome duplication" (2R hypothesis) hypothesis was made and gained its popularity since then, especially among developmental biologists.

However, the 2R hypothesis has also been questioned by many researchers. Based on the theory, human genome should have a 4:1 ratio compared with invertebrate's genome. This is not supported by recent finding from various genome projects — human genome consists of ~35,000 genes while an average invertebrate genome sized about 15,000 genes. Diploidization could not fully justify this ratio because phylogenetic studies failed to resolve a four-gene-clusters of (AB) (CD) tree topology in vertebrates.

The 2R hypothesis is still under debate today, with supporters on both sides. Some critical evidence might be found in the root of vertebrate lineages, such as *Ciona intestinalis*. With more and more ESTs available, we soon would have a better understanding on the paleopolyploidy of ourselves.

SYSTEMS BIOLOGY

Systems biology, a field of study in the biosciences, focuses on the systematic study of complex interactions in biological systems. Particularly from 2000 onwards, the term is used widely in the biosciences, and in a variety of contexts.

Overview

Systems biology can be considered from a number of different aspects:

- Some sources discuss systems biology as a field of study, particularly, the study of the interactions between the components of *biological systems*, and how these interactions give rise to the function and behavior of that system.
- Other sources consider systems biology as a paradigm, usually defined in antithesis to the so-called reductionist paradigm, although fully consistent with the scientific method. The distinction between the two paradigms is referred to in these quotations:

  "The reductionist approach has successfully identified most of the components and many of the interactions but, unfortunately, offers no convincing concepts or methods to understand how system properties emerge...the pluralism of causes and effects in biological networks is better addressed by observing, through quantitative measures, multiple components simultaneously and by rigorous data integration with mathematical models" Science

  "Systems biology...is about putting together rather than taking apart, integration rather than reduction. It requires that we develop ways of thinking about integration that are as rigorous as our reductionist programmes, but different....It means changing our philosophy, in the full sense of the term" Denis Noble

- Still other sources view systems biology in terms of the operational protocols used for performing research, namely a cycle composed of theory, computational modelling to propose specific testable hypotheses about a biological system, experimental validation, and then using the newly acquired quantitative description of cells or cell processes to refine the computational model or theory.. Since the objective is a model of the interactions in a system, the experimental techniques that most suit systems biology are those that are system-wide and attempt to be as complete as possible. Therefore, transcriptomics, metabolomics, proteomics and high-throughput techniques are used to collect quantitative data for the construction and validation of models.
- Finally, some sources see it as a socioscientific phenomenon defined by the strategy of pursuing integration of complex data about the interactions in biological systems from diverse experimental sources using interdisciplinary tools and personnel.

This variety of viewpoints is illustrative of the fact that systems biology refers to a cluster of peripherally overlapping concepts rather than a single well-delineated field. However the term has widespread currency and popularity as of 2007, with chairs and institutes of systems biology proliferating worldwide.

History

Systems biology finds its roots in:

- the quantitative modelling of enzyme kinetics, a discipline that flourished between 1900 and 1970,
- the simulations developed to study neurophysiology, and
- control theory, and cybernetics.

One of the theorists who can be seen as a precursor of systems biology is Ludwig von Bertalanffy with his general systems theory. In 1952, the British neurophysiologists and nobel prize winners Alan Lloyd Hodgkin and Andrew Fielding Huxley constructed a mathematical model describing the action potential propagating along the axon of a neuronal cell. In 1960, Denis Noble developed the first computer model of a beating heart.

The 1960s and 1970s saw the development of several approaches to study complex molecular systems, such as the Metabolic Control Analysis and the biochemical systems theory. The successes of molecular biology throughout the 1980s, coupled with a skepticism toward theoretical biology, that then promised more than it achieved, caused the quantitative modelling of biological processes to become a somewhat minor field.

However the birth of functional genomics in the 1990s meant that large quantities of high quality data became available, while the computing power exploded, making more realistic models possible. In 1997, the group of Masaru Tomita published the first quantitative model of the metabolism of a whole (hypothetical) cell.

Around the year 2000, when Institutes of Systems Biology were established in Seattle and Tokyo, systems biology emerged as a movement in its own right, spurred on by the completion of various genome projects, the large increase in data from the omics (e.g. genomics and proteomics) and the accompanying advances in high-throughput experiments and bioinformatics. Since then, various research institutes dedicated to systems biology have been developed. As of summer 2006, due to a shortage of people in systems biology several doctoral training centres in systems biology have been established in many parts of the world.

Techniques Associated with Systems Biology

According to the interpretation of System Biology as the ability to obtain, integrate and analyze complex data from multiple experimental sources using interdisciplinary tools, some typical technology platforms are:

- Transcriptomics: whole cell or tissue gene expression measurements by DNA microarrays or SAGE
- Proteomics: complete identification of proteins and protein expression patterns of a cell or tissue through two-dimensional gel electrophoresis and mass spectrometry or multi-dimensional protein identification techniques (advanced HPLC systems coupled with mass spectrometry). Sub disciplines include phosphoproteomics, glycoproteomics and other methods to detect chemically modified proteins.
- Metabolomics: identification and measurement of all small-molecules metabolites within a cell or tissue
- Glycomics: identification of the entirety of all carbohydrates in a cell or tissue.

In addition to the identification and quantification of the above given molecules further techniques analyze the dynamics and interactions within a cell. This includes:

- Interactomics which is used mostly in the context of protein-protein interaction but in theory encompasses interactions between all molecules within a cell
- Fluxomics, which deals with the dynamic changes of molecules within a cell over time

The investigations are frequently combined with large scale perturbation methods, including gene-based (RNAi, mis-expression of wild type and mutant genes) and chemical approaches using small molecule libraries. Robots and automated sensors enable such large-scale experimentation and data acquisition. These technologies are still emerging and many face problems that the larger the quantity of data produced, the lower the quality. A wide variety of quantitative scientists (computational biologists, statisticians, mathematicians, computer scientists, engineers, and physicists) are working to improve the quality of these approaches and to create, refine, and retest the models to accurately reflect observations.

The investigations of a single level of biological organization (such as those listed above) are usually referred to as Systematic Systems Biology. Other areas of Systems Biology includes Integrative Systems Biology, which seeks to integrate different types of information to advance the understanding the biological whole, and Dynamic Systems Biology, which aims to uncover how the biological whole changes over time (during evolution, for example, the onset of disease or in response to a perturbation). Functional Genomics may also be considered a sub-field of Systems Biology.

The systems biology approach often involves the development of mechanistic models, such as the reconstruction of dynamic systems from the quantitative properties of their elementary building blocks. For instance, a cellular network can be modelled mathematically using methods coming from chemical kinetics and control theory. Due to the large number of parameters, variables and constraints in cellular networks, numerical and computational techniques are often used. Other aspects of computer science and informatics are also used in systems biology. These include new forms of computational model, such as the use of process calculi to model biological processes, the integration of information from the literature, using techniques of information extraction and text mining, the development of online databases and repositories for sharing data and models (such as BioModels Database), approaches to database integration and software interoperability via loose coupling of software, websites and databases (such as Gaggle), and the development of syntactically and semantically sound ways of representing biological models, such as the Systems Biology Markup Language.

GENOME PROJECT

Genome projects are scientific endeavours that ultimately aim to determine the complete genome sequence of an organism (be it an animal, a plant, a fungus, a bacterium, an archaean, a protist or a virus). The genome sequence for any organism requires the DNA sequences for each of the chromosomes in an organism to be determined. For bacteria, which usually have just one chromosome, a genome project will aim to map the sequence of that chromosome. Humans, with 22 pairs of chromosomes and 2 sex chromosomes, will require 24 separate chromosome sequences in order to represent the completed genome.

The Human Genome Project was a landmark genome project and some have argued that the era of genomics is one of the more fundamental advances in human history.

Genome Assembly

Genome assembly refers to the process of taking a large number of short DNA sequences, all of which were generated by a shotgun sequencing project, and putting them back together to create a representation of the original chromosomes from which the DNA originated. In a shotgun sequencing project, all the DNA from a source (usually a single organism, anything from a bacterium to a mammal) is first fractured into millions of small pieces. These pieces are then "read" by automated sequencing machines, which can read up to 900 nucleotides or bases at a time. (The four bases are adenine, guanine, cytosine, and thymine, represented as AGCT.) A genome assembly algorithm works by taking all the pieces and aligning them to one another, and detecting all places where two of the short sequences, or *reads*, overlap. These overlapping reads can be merged together, and the process continues.

Genome assembly is a very difficult computational problem, made more difficult because genomes contain large numbers of identical sequences, known as *repeats*. These repeats can be thousands of nucleotides long, and some occur in thousands of different locations, especially in the large genomes of plants and animals.

Assembly Software

Most research institutes that sequence DNA use their own software for assembling the sequences that they produce. Some well known assembly programs include:

Phred/Phrap by Phil Green

AMOS (A Modular, Open-Source assembler) is a well-known open source effort to bring together the efforts of leading genome assembly code developers. The home of AMOS is currently http://amos.sourceforge.net. AMOS was initiated at The Institute for Genomic Research by Steven Salzberg, Mihai Pop, and Art Delcher.

The Celera Assembler was the assembler developed by Gene Myers, Granger Sutton, Art Delcher, and others at Celera Genomics from 1998 until approximately 2002. It was moved to SourceForge and continues to be developed by the original scientists and others, at http://sourceforge.net/projects/wgs-assembler.

The Arachne assembler began in 2000 as the doctoral thesis of Serafim Batzoglou, now at Stanford University. Since that time, it has been developed by a team lead by David B. Jaffe at the Broad Institute, formerly part of the Whitehead Institute. It is available for download at http://www.broad.mit.edu/wga/.

Genome Annotation

Genome annotation is the process of attaching biological information to sequences. It consists of two main steps:

1. identifying elements on the genome, a process called Gene Finding, and
2. attaching biological information to these elements.

Automatic annotation tools try to perform all this by computer analysis, as opposed to manual annotation which involves human expertise. Ideally, these approaches co-exist and complement each other in the same annotation pipeline.

The basic level of annotation is using BLAST for finding similarities, and then annotating genomes based on that. However, nowadays more and more additional information is added to the annotation platform. The additional information allows manual annotators to deconvolute discrepancies between genes that are given the same annotation.

For example, the SEED database uses genome context information, similarity scores, experimental data, and integrations of other resources to provide the most accurate genome annotations through their Subsystems approach.

Structural annotation consists in the identification of genomic elements.

- ORFs and their localisation
- gene structure
- coding regions
- location of regulatory motifs

Functional annotation consists in attaching biological information to genomic elements.

- biochemical function
- biological function
- involved regulation and interactions
- expression

These steps may involve both biological experiments and *in silico* analysis.

A variety of software tools have been developed to permit scientists to view and share genome annotations.

When is a Genome Project Finished?

When sequencing a genome, there are usually regions that are difficult to sequence (often regions with highly repetitive DNA). Thus, 'completed' genome sequences are rarely ever complete, and terms such as 'working draft' or 'essentially complete' have been used to more accurately describe the status of such genome projects. Even when every base pair of a genome sequence has been determined, there are still likely to be errors present because DNA sequencing is not a completely accurate process. It could also be argued that a complete genome project should include the sequences of mitochondria and (for plants) chloroplasts as these organelles have their own genomes.

It is often reported that the goal of sequencing a genome is to obtain information about the complete set of genes in that particular genome sequence. The proportion of a genome that encodes for genes may be very small (particularly in eukaryotes such as humans, where coding DNA may only account for a few percent of the entire sequence). However, it is not always possible (or desirable) to only sequence the coding regions separately. Also, as scientists understand more about the role of this noncoding DNA (often referred to as junk DNA), it will become more important to have a complete genome sequence as a background to understanding the genetics and biology of any given organism.

In many ways genome projects do not confine themselves to only determining a DNA sequence of an organism. Such projects may also include gene prediction to find out where the genes are in a genome, and what those genes do. There may also be related projects to sequence ESTs or mRNAs to help find out where the genes actually are.

Historical and Technological Perspectives

Historically, when sequencing eukaryotic genomes (such as the worm *Caenorhabditis elegans*) it was common to first map the genome to provide a series of landmarks across the genome. Rather than sequence a chromosome in one go, it would be sequenced piece by piece (with the prior knowledge of approximately where that piece is located on the larger chromosome). Changes in technology and in particular improvements to the processing power of computers, means that genomes can now be 'shotgun sequenced' in one go (there are caveats to this approach though when compared to the traditional approach).

Improvements in DNA sequencing technology has meant that the cost of sequencing a new genome sequence has steadily fallen (in terms of cost per base pair) and newer technology has also meant that genomes can be sequenced far more quickly. When research agencies decide what new genomes to sequence, the emphasis has been on species which have either a relevance to human health (e.g. pathogenic bacteria or vectors of disease such as mosquitos) or species which have commercial importance (e.g. livestock and crop plants). Secondary emphasis is placed on species whose genomes will help answer important questions in molecular evolution (e.g. the common chimpanzee).

In the future, it is likely that it will become even cheaper and quicker to sequence a genome. This will allow for complete genome sequences to be determined from many different individuals of the same species. For humans, this will allow us to better understand aspects of human genetic diversity.

Example Genome Projects

Many organisms have genome projects that have either been completed or will be completed shortly, including:

- Humans, *Homo sapiens*; see Human genome project
- Neanderthal, "*Homo neanderthalensis*" (partial)

- *Haemophilus influenzae*, a bacterium (the first free-living organism to have its genome fully sequenced)
- Common House Mouse, *Mus musculus*
- Brown Rat, *Rattus norvegicus*
- Common Chimpanzee *Pan troglodytes*; see Chimpanzee Genome Project
- Rhesus Macaque, *Macaca mulatta*
- Domestic Chicken, *Gallus gallus*
- Tammar Wallaby, *Macropus eugenii*
- Domestic Cat, *Felis silvestris*
- Domestic Dog, *Canis lupus familiaris*
- Common fruit fly, *Drosophila melanogaster*
- Baker's yeast, *Saccharomyces cerevisiae*
- Red bread mold, *Neurospora crassa*
- Thale Cress, *Arabidopsis thaliana*
- Rice, *Oryza sativa*
- Common Wheat, *Triticum aestivum*
- Maize, *Zea mays*
- Poplar, Populus trichocarpa (The first tree to have its genome fully sequenced)
- *Escherichia coli*, a coliform bacterium
- SARS virus
- Purple-spined sea urchin, *Arbacia punctulata*
- *Caenorhabditis elegans*, a nematode worm
- Zebra Danio, *Brachydanio rerio*
- African clawed frog, *Xenopus laevis*
- *Oryzias latipes*, a medakafish
- Tiger blowfish, *Takifugu rubipres*
- Tomato *Solanum lycopersicum*
- Potato *Solanum tuberosum*
- Western Honey bee, *Apis mellifera*
- Grapevine, *Vitis vinifera L.*
- Spanish flu

HUMAN GENOME PROJECT

The Human Genome Project (HGP) is an international scientific research project. Its goal is to understand the genetic make-up of the human species by determining the sequence of chemical base pairs which make up DNA, and to identify the 20,000-25,000 genes of the human genome. The project also focused on several other nonhuman organisms such as Escherichia coli, the fruit fly, and a laboratory mouse. It was one of the largest investigational projects in modern science.

The project began in 1990 initially headed by James D. Watson. A working draft of the genome was released in 2000 and a complete one in 2003, with further analysis still being published.

A parallel project was conducted by the private company Celera Genomics. Most of the sequencing was performed in universities and research centers from the United States and Great Britain.

The mapping of human genes is an important step in the development of medicines and other aspects of health care.

The HGP originally aimed to map the nucleotides contained in a haploid reference human genome (more than three billion). Several groups have announced efforts to extend this to diploid human genomes including the International HapMap Project, Applied Biosystems, Perlegen, Illumina, JCVI, Personal Genome Project, and Roche-454.

The "genome" of any given individual (except for identical twins and cloned animals) is unique; mapping "the human genome" involves sequencing multiple variations of each gene The project did not study all of the DNA found in human cells; some heterochromatic areas (about 8% of the total) remain un-sequenced.

The Project

Background

Initiation of the Project was the culmination of several years of work supported by the Department of Energy, in particular workshops in 1984 and 1986 and a subsequent initiative the Department of Energy. This 1986 report stated boldly, "The ultimate goal of this initiative is to understand the human genome" and "Knowledge of the human genome is as necessary to the continuing progress of medicine and other health sciences as knowledge of human anatomy has been for the present state of medicine." Candidate technologies were already being considered for the proposed undertaking at least as early as 1985.

James D. Watson was Head of the National Center for Human Genome Research at the National Institutes of Health (NIH) in the United States starting from 1988. Largely due to his disagreement with his boss, Bernadine Healy, over the issue of patenting genes, he was forced to resign in 1992. He was replaced by Francis Collins in April 1993, and the name of the Center was changed to the National Human Genome Research Institute (NHGRI) in 1997.

The $3-billion project was formally founded in 1990 by the United States Department of Energy and the U.S. National Institutes of Health, and was expected to take 15 years. In addition to the United States, the international consortium comprised geneticists in China, France, Germany, Japan, and the United Kingdom.

Due to widespread international cooperation and advances in the field of genomics (especially in sequence analysis), as well as major advances in computing technology, a 'rough draft' of the genome was finished in 2000 (announced jointly by then US president Bill Clinton and British Prime Minister Tony Blair on June 26, 2000). Ongoing sequencing led to the announcement of the essentially complete genome in April 2003, 2 years earlier than planned. In May 2006, another milestone was passed on the way to completion of the project, when the sequence of the last chromosome was published in the journal Nature.

There are multiple definitions of the "complete sequence of the human genome". According to some of these definitions, the genome has already been completely sequenced, and according to other definitions, the genome has yet to be completely sequenced. There have been multiple popular press articles reporting that the genome was "complete." The genome has been completely sequenced using the definition employed by the International Human Genome Project. A graphical history of the human genome project shows that most of the human genome was complete by the end of 2003. However, there are a number of regions of the human genome that can be considered unfinished. First, the central regions of each chromosome, known as centromeres, are highly repetitive DNA sequences that are difficult to sequence using current technology. The centromeres are millions (possibly tens of millions) of base pairs long, and for the most part these are entirely un-sequenced. Second, the ends of the chromosomes, called telomeres, are also highly repetitive, and for most of the 46 chromosome ends these too are incomplete. We do not know precisely how much sequence remains before we reach the telomeres of each chromosome, but as with the centromeres, current technology does not make it easy to get there. Third, there are several loci in each individual's genome that contain members of multigene families that are difficult to disentangle with shotgun sequencing methodologies - these multigene families often encode proteins important for immune functions. It is likely that the centromeres and telomeres will remain un-sequenced until new technology is developed that facilitates their sequencing. Other than these regions, there remain a few dozen gaps scattered around the genome, some of them rather large, but there is hope that all these will be closed in the next couple of years. In summary: our best estimates of total genome size indicate that about 92% of the genome has been completed . Most of the remaining DNA is highly repetitive and unlikely to contain genes, but we cannot truly know until we sequence all of it. Understanding the functions of all the genes and their regulation is far from complete. The roles of junk DNA, the evolution of the genome, the differences between individuals, and many other questions are still the subject of intense study by laboratories all over the world.

Goals

The goals of the original HGP were not only to determine more than 3 billion base pairs in the human genome with a minimal error rate, but also to identify all the genes in this vast amount of data. This part of the project is still ongoing, although a preliminary count indicates about 22,000–23,000 genes in the human genome, which is fewer than predicted by many scientists.

Another goal of the HGP was to develop faster, more efficient methods for DNA sequencing and sequence analysis and the transfer of these technologies to industry.

The sequence of the human DNA is stored in databases available to anyone on the Internet. The U.S. National Center for Biotechnology Information (and sister organizations in Europe and Japan) house the gene sequence in a database known as Genbank, along with sequences of known and hypothetical genes and proteins. Other organizations such as the University of California, Santa Cruz, and Ensembl present additional data and annotation and powerful tools for visualizing and searching it. Computer programs have been developed to analyze the data, because the data themselves are difficult to interpret without such programs.

The process of identifying the boundaries between genes and other features in raw DNA sequence is called genome annotation and is the domain of bioinformatics. While expert biologists make the best annotators, their work proceeds slowly, and computer programs are increasingly used to meet the high-throughput demands of genome sequencing projects. The best current technologies for annotation make use of statistical models that take advantage of parallels between DNA sequences and human language, using concepts from computer science such as formal grammars.

Another, often overlooked, goal of the HGP is the study of its ethical, legal, and social implications. It is important to research these issues and find the most appropriate solutions before they become large dilemmas whose effect will manifest in the form of major political concerns.

All humans have unique gene sequences; therefore the data published by the HGP does not represent the exact sequence of each and every individual's genome. It is the combined genome of a small number of anonymous donors. The HGP genome is a scaffold for future work in identifying differences among individuals. Most of the current effort in identifying differences among individuals involves single nucleotide polymorphisms and the HapMap.

How it was Accomplished

Funding came from the US government through the National Institutes of Health in the United States, and the UK charity, the Wellcome Trust, who funded the Sanger Institute (then the Sanger Centre) in Great Britain, as well as numerous other groups from around the world. The genome was broken into smaller pieces; approximately 150,000 base pairs in length. These pieces are called "bacterial artificial chromosomes", or BACs, because they can be inserted into bacteria where they are copied by the bacterial DNA replication machinery. Each of these pieces was then sequenced separately as a small "shotgun" project and then assembled. The larger, 150,000 base pairs go together to create chromosomes. This is known as the "hierarchical shotgun" approach, because the genome is first broken into relatively large chunks, which are then mapped to chromosomes before being selected for sequencing.

Celera Genomics HGP

In 1998, a similar, privately funded quest was launched by the American researcher Craig Venter and his firm Celera Genomics. The $300 million Celera effort was intended to proceed at a faster pace and at a fraction of the cost of the roughly $3 billion publicly funded project.

Celera used a riskier technique called whole genome shotgun sequencing, which had been used to sequence bacterial genomes of up to six million base pairs in length, but not for anything nearly as large as the three thousand million base pair human genome.

Celera initially announced that it would seek patent protection on "only 200-300" genes, but later amended this to seeking "intellectual property protection" on "fully-characterized important structures" amounting to 100-300 targets. The firm eventually filed preliminary ("place-holder") patent applications on 6,500 whole or partial genes. Celera also promised to publish their findings in accordance with the terms of the 1996 "Bermuda Statement," by releasing new data quarterly (the HGP released its new data daily), although, unlike the publicly funded project, they would not permit free redistribution or commercial use of the data.

In March 2000, President Clinton announced that the genome sequence could not be patented, and should be made freely available to all researchers. The statement sent Celera's stock plummeting and dragged down the biotechnology-heavy Nasdaq. The biotechnology sector lost about $50 billion in market capitalization in two days.

Although the working draft was announced in June 2000, it was not until February 2001 that Celera and the HGP scientists published details of their drafts. Special issues of *Nature* (which published the publicly funded project's scientific paper) and *Science* (which published Celera's paper) described the methods used to produce the draft sequence and offered analysis of the sequence. These drafts covered about 83% of the genome (90% of the euchromatic regions with 150,000 gaps and the order and orientation of many segments not yet established). In February 2001, at the time of the joint publications, press releases announced that the project had been completed by both groups. Improved drafts were announced in 2003 and 2005, filling in to ~92% of the sequence currently.

The competition proved to be very good for the project, spurring the public groups to modify their strategy in order to accelerate progress. The rivals initially agreed to pool their data, but the agreement fell apart when Celera refused to deposit its data in the unrestricted public database GenBank. Celera had incorporated the public data into their genome, but forbade the public effort to use Celera data.

HGP is the most well known of many international genome projects aimed at sequencing the DNA of a specific organism. While the human DNA sequence offers the most tangible benefits, important developments in biology and medicine are predicted as a result of the sequencing of model organisms, including mice, fruit flies, zebrafish, yeast, nematodes, plants, and many microbial organisms and parasites.

In 2004, researchers from the International Human Genome Sequencing Consortium (IHGSC) of the HGP announced a new estimate of 20,000 to 25,000 genes in the human genome. Previously 30,000 to 40,000 had been predicted, while estimates at the start of the project reached up to as high as 2,000,000. The number continues to fluctuate and it is now expected that it will take many years to agree on a precise value for the number of genes in the human genome.

History

In 1976, the genome of the virus Bacteriophage MS2 was the first complete genome to be determined, by Walter Fiers and his team at the University of Ghent (Ghent, Belgium). The idea for the shotgun technique came from the use of an algorithm that combined sequence information from many small fragments of DNA to reconstruct a genome. This technique was pioneered by Frederick Sanger to sequence the genome of the Phage Ö-X174, a tiny virus called a bacteriophage that was the first fully sequenced genome (DNA-sequence) in 1977. The technique was called shotgun sequencing because the genome was broken into millions of pieces as if it had been blasted with a shotgun. In order to scale up the method, both the sequencing and genome assembly had to be automated, as they were in the 1980s.

Those techniques were shown applicable to sequencing of the first free-living bacterial genome (1.8 million base pairs) of *Haemophilus influenzae* in 1995 and the first animal genome (~100 Mbp) It involved the use of automated sequencers, longer individual sequences using approximately 500 base pairs at that time. Paired sequences separated by a fixed distance of around 2000 base pairs which were critical elements enabling the development of the first genome assembly programs for reconstruction of large regions of genomes (aka 'contigs').

Three years later, in 1998, the announcement by the newly-formed Celera Genomics that it would sçale up the shotgun sequencing method to the human genome was greeted with skepticism in some circles. The shotgun technique breaks the DNA into fragments of various sizes, ranging from 2,000 to 300,000 base pairs in length, forming what is called a DNA "library". Using an automated DNA sequencer the DNA is read in 800bp lengths from both ends of each fragment. Using a complex genome assembly algorithm and a supercomputer, the pieces are combined and the genome can be reconstructed from the millions of short, 800 base pair fragments. The success of both the public and privately funded effort hinged upon a new, more highly automated capillary DNA sequencing machine, called the Applied Biosystems 3700, that ran the DNA sequences through an extremely fine capillary tube rather than a flat gel. Even more critical was the development of a new, larger-scale genome assembly program, which could handle the 30-50 million sequences that would be required to sequence the entire human genome with this method. At the time, such a program did not exist. One of the first major projects at Celera Genomics was the development of this assembler, which was written in parallel with the construction of a large, highly automated genome sequencing factory. Development of the assembler was led by Gene Myers. The first version of this assembler was demonstrated in 2000, when the Celera team joined forces with Professor Gerald Rubin to sequence the fruit fly Drosophila melanogaster using the whole-genome shotgun method. At 130 million base pairs, it was at least 10 times larger than any genome previously shotgun assembled. One year later, the Celera team published their assembly of the three billion base pair human genome.

How it was Accomplished

The IHGSC used pair-end sequencing plus whole-genome shotgun mapping of large (~100 Kbp) plasmid clones and shotgun sequencing of smaller plasmid sub-clones plus a variety of other mapping data to orient and check the assembly of each human chromosome.

The Celera group tried "whole-genome shotgun" sequencing without using the additional mapping scaffolding, but by including shredded public data raised questions.

Whose Genome was Sequenced?

In the IHGSC international public-sector Human Genome Project (HGP), researchers collected blood (female) or sperm (male) samples from a large number of donors. Only a few of many collected samples were processed as DNA resources. Thus the donor identities were protected so neither donors nor scientists could know whose DNA was sequenced. DNA clones from many different libraries were used in the overall project, with most of those libraries being created by Dr. Pieter J. de Jong. It has been informally reported, and is well known in the genomics community, that much of the DNA for the public HGP came from a single anonymous male donor from Buffalo, New York (code name RP11).

HGP scientists used white blood cells from the blood of 2 male and 2 female donors (randomly selected from 20 of each) — each donor yielding a separate DNA library. One of these libraries (RP11) was used considerably more than others, due to quality considerations. One minor technical issue is that male samples contain only half as much DNA from the X and Y chromosomes as from the other 22 chromosomes (the autosomes); this happens because each male cell contains only one X and one Y chromosome, not two like other chromosomes (autosomes). (This is true for nearly all male cells, not just sperm cells.)

Although the main sequencing phase of the HGP has been completed, studies of DNA variation continue in the International HapMap Project, whose goal is to identify patterns of single nucleotide polymorphism (SNP) groups (called haplotypes, or "haps"). The DNA samples for the HapMap came from a total of 270 individuals: Yoruba people in Ibadan, Nigeria; Japanese people in Tokyo; Han Chinese in Beijing; and the French Centre d'Etude du Polymorphisms Humain (CEPH) resource, which consisted of residents of the United States having ancestry from Western and Northern Europe.

In the Celera Genomics private-sector project, DNA from five different individuals were used for sequencing. The lead scientist of Celera Genomics at that time, Craig Venter, later acknowledged (in a public letter to the journal Science) that his DNA was one of those in the pool.

On September 4th, 2007, a team led by Craig Venter, published his complete DNA sequence, unveiling the six-billion-letter genome of a single individual for the first time.

Benefit

The work on interpretation of genome data is still in its initial stages. It is anticipated that detailed knowledge of the human genome will provide new avenues for advances in medicine and bio-technology. Clear practical results of the project emerged even before the work was finished. For example, a number of companies, such as Myriad Genetics started offering easy ways to administer genetic tests that can show predisposition to a variety of illnesses, including breast cancer, disorders of hemostasis, cystic fibrosis, liver diseases and many others. Also, the etiologies for cancers, Alzheimer's disease and other areas of clinical interest are considered likely to benefit from genome information and possibly may lead in the long term to significant advances in their management.

There are also many tangible benefits for biological scientists. For example, a researcher investigating a certain form of cancer may have narrowed down his/her search to a particular gene. By visiting the human genome database on the worldwide web, this researcher can examine what other scientists have written about this gene, including (potentially) the three-dimensional structure of its product, its function(s), its evolutionary relationships to other human genes, or to genes in mice or yeast or fruit flies, possible detrimental mutations, interactions with other genes, body tissues in which this gene is activated, diseases associated with this gene or other datatypes.

Further, deeper understanding of the disease processes at the level of molecular biology may determine new therapeutic procedures. Given the established importance of DNA in molecular biology and its central role in determining the fundamental operation of cellular processes, it is likely that expanded knowledge in this area will facilitate medical advances in numerous areas of clinical interest that may not have been possible without them.

The analysis of similarities between DNA sequences from different organisms is also opening new avenues in the study of the theory of evolution. In many cases, evolutionary questions can now be framed in terms of molecular biology; indeed, many major evolutionary milestones (the emergence of the ribosome and organelles, the development of embryos with body plans, the vertebrate immune system) can be related to the molecular level. Many questions about the similarities and differences between humans and our closest relatives (the primates, and indeed the other mammals) are expected to be illuminated by the data from this project.

The Human Genome Diversity Project, spinoff research aimed at mapping the DNA that varies between human ethnic groups, which was rumored to have been halted, actually did continue and to date has yielded new conclusions. In the future, HGDP could possibly expose new data in disease surveillance, human development and anthropology. HGDP could unlock secrets behind and create new strategies for managing the vulnerability of ethnic groups to certain diseases (see race in biomedicine). It could also show how human populations have adapted to these vulnerabilities.

2R HYPOTHESIS

The 2R hypothesis or Ohno's hypothesis, first proposed by Susumu Ohno in 1970, is a contested hypothesis in genomics and molecular evolution suggesting that the genomes of the early vertebrate lineage underwent one or more complete genome duplications, and thus modern vertebrate genomes reflect paleopolyploidy. The name derives from the *2 rounds* of duplication hypothesized by a 1994 version, and the term *2R hypothesis* was probably coined in 1999; variations in the number of duplications typically still are referred to as examples of the 2R hypothesis. Since Ohno proposed the first version of it in *Evolution by Gene Duplication*, the 2R hypothesis has been the subject of much research, but even with recent data from the human genome, it remains a matter of scientific dispute.

Ohno's Argument

Ohno presented the first version of the 2R hypothesis as part of his larger argument for the general importance of gene duplication in evolution. Based on relative genome sizes, he suggested that ancestral fish or amphibians had undergone at least one and possibly more cases of "tetraploid evolution". He later added to this argument the evidence that most paralogous genes in vertebrates do not demonstrate genetic linkage. Ohno argued that linkage should be expected in the case of individual tandem duplications (in which a duplicate gene is added adjacent to the original gene on the same chromosome), but not in the case of chromosome duplications.

Later Evidence

The 2R hypothesis saw a resurgence of interest in the 1990s, with multiple suggestions for the time and number of duplications. Estimates for the dates of duplication events range from before 640 million years ago to after 450 million years ago. One argument for the hypothesis relies on the number of genes in mammals, which was previously estimated at around four times the number in invertebrate genomes (though more recent estimates are less than half that). Some analyses of gene families in human chromosomes have shown patterns that are not consistent with the 2R hypothesis. Parsimony analysis has produced some results that, while not supportive of the hypothesis, do not rule it out. According to a 2001 review of the subject by Wojciech Maka³owski, "the hypothesis of whole genome duplications in the early stages of vertebrate evolution has as many adherents as opponents"; Maka³owski argues that the overall balance of current evidence seems to be against it, but that it is not easily falsifiable because it can be modified to accommodate a wide range of data. Others analyzing duplications in the draft human genome sequence have claimed that genome data provides evidence of "extensive duplication" and that the parsimony tests that cast doubt on the hypothesis are of questionable validity. A review in 2007 by Masanori Kasahara states that there is now "incontrovertible evidence supporting the 2R hypothesis" and that "a long-standing debate on the 2R hypothesis is approaching the end".

REFERENCES

A. Wagner (1997), A computational genomics approach to the identification of gene networks, Nucleic Acids Res., Sep 15;25(18):3594-604,

Adams KL, Wendel JF., Polyploidy and genome evolution in plants. Curr Opin Plant Biol. 2005 Apr;8(2):135-41. Review. PubMed

Adams, MD. et al. (2000). "The genome sequence of Drosophila melanogaster.". Science 287: 2185"2195.

Barnhart, Benjamin J. (1989). "DOE Human Genome Program". Human Genome Quarterly 1: 1. Retrieved 2005-02-03.

Blanc G, Wolfe KH. Functional divergence of duplicated genes formed by polyploidy during Arabidopsis evolution. Plant Cell. 2004 Jul;16(7):1679-91. Epub 2004 June 18. PubMed

Blanc G, Wolfe KH. Widespread paleopolyploidy in model plant species inferred from age distributions of duplicate genes. Plant Cell. 2004 Jul;16(7):1667-78. Epub 2004 June 18. PubMed

C. elegans Sequencing Consortium (1998). "Genome sequence of the nematode C. elegans: A platform for investigating biology.". Science 282: 2012–18.

Comai L. The advantages and disadvantages of being polyploid. Nat Rev Genet. 2005 Nov;6(11):836-46. Review. PubMed

Cristianini, N. and Hahn, M. Introduction to Computational Genomics, Cambridge University Press, 2006.

Cui L, Wall PK, Leebens-Mack JH, Lindsay BG, Soltis DE, Doyle JJ, Soltis PS, Carlson JE, Arumuganathan K, Barakat A, Albert VA, Ma H, dePamphilis CW. Widespread genome duplications throughout the history of flowering plants. Genome Res. 2006 Jun;16(6):738-49. Epub 2006 May 15. PubMed

David Mount (2000), Bioinformatics, Sequence and Genome Analysis, pp. 2-3, Cold Spring Harbor Laboratory Press,

DeLisi, Charles (2001). "Genomes: 15 Years Later A Perspective by Charles DeLisi, HGP Pioneer". Human Genome News 11: 3–4. Retrieved 2005-02-03.

Denis Noble The Music of Life Oxford University Press (2006) p21

Fiers W et al., Complete nucleotide-sequence of bacteriophage MS2-RNA - primary and secondary structure of replicase gene, Nature, 260, 500-507, 1976

Fiers W, Contreres R, Duerinck F, Haegeman G, Iserentant D, Merregaert J, Min Jou W, Molemans F, Raeymaekers A, Van den Berghe A, Volckaert G, Ysebaert M. Complete nucleotide sequence of bacteriophage MS2 RNA: primary and secondary structure of the replicase gene, Nature. 1976 Apr 8;260(5551):500-7.

Fleischmann, R. D. et al. (1995). "Whole-genome random sequencing and assembly of Haemophilus influenzae Rd.". Science 269: 496"512.

Herrero A and Flores E (editor). (2008). The Cyanobacteria: Molecular Biology, Genomics and Evolution, 1st ed., Caister Academic Press.

IHGSC (2004). "Finishing the euchromatic sequence of the human genome.". Nature 431: 931-945.

International Human Genome Sequencing Consortium (2001). "Initial sequencing and analysis of the human genome.". Nature 409: 860"921.

Kellis M, Birren BW, Lander ES., Proof and evolutionary analysis of ancient genome duplication in the yeast Saccharomyces cerevisiae., Nature. 2004 April 8;428(6983):617-24. Epub 2004 March 7. PubMed

Kennedy D (2002). "Not wicked, perhaps, but tacky". Science 297: 1237.

Kholodenko B.N., Bruggeman F.J., Sauro H.M.; Alberghina L. and Westerhoff H.V.(Eds.) (2005.). "Mechanistic and modular approaches to modeling and inference of cellular regulatory networks". Systems Biology: Definitions and Perspectives: p143, Springer-Verlag.

Koonin EV (2001) Computational Genomics, National Center for Biotechnology Information, National Library of Medicine, NIH (PubMed ID: 11267880)

Levy S, Sutton G, Ng PC, Feuk L, Halpern AL, et al. (2007). "The Diploid Genome Sequence of an Individual Human". PLoS Biology 5 (10).

Mc Grath S and van Sinderen D (editors). (2007). Bacteriophage: Genetics and Molecular Biology, 1st ed., Caister Academic Press.

Min Jou W, Haegeman G, Ysebaert M, Fiers W., Nucleotide sequence of the gene coding for the bacteriophage MS2 coat protein, Nature. 1972 May 12;237(5350):82-8

Osoegawa, Kazutoyo (2001). "A Bacterial Artificial Chromosome Library for Sequencing the Complete Human Genome". Genome Research 11: 483-496.

Otto SP, Whitton J., Polyploid incidence and evolution., Annu Rev Genet. 2000;34:401-437. PubMed

Sanger F, Air GM, Barrell BG, Brown NL, Coulson AR, Fiddes CA, Hutchison CA, Slocombe PM, Smith M., Nucleotide sequence of bacteriophage phi X174 DNA, Nature. 1977 Feb 24;265(5596):687-95

Sanger F, Air GM, Barrell BG, Brown NL, Coulson AR, Fiddes CA, Hutchison CA, Slocombe PM, Smith M., Nucleotide sequence of bacteriophage phi X174 DNA, Nature. 1977 Feb 24;265(5596):687-95

Sauer, U. et al. "Getting Closer to the Whole Picture" Science (journal) 316 550 17 April 2007

Snoep J.L. and Westerhoff H.V.; Alberghina L. and Westerhoff H.V. (Eds.) (2005.). "From isolation to integration, a systems biology approach for building the Silicon Cell". Systems Biology: Definitions and Perspectives: p7, Springer-Verlag.

Venter, JC, et al (2001). "The sequence of the human genome.". Science 291: 1304"1351.

Waterston RH, Lander ES, Sulston JE (2003). "More on the sequencing of the human genome". Proc Natl Acad Sci U S A. 100: 3022-4.

Werner, E., "All systems go"., Naturevol 446, pp 493-494, March 29, 2007.

Wojciech Maka³owski, Are We Polyploids? A Brief History of One Hypothesis., 2001 11: 667-670 Genome Res. pdf

Wolfe KH. Related, Yesterday's polyploids and the mystery of diploidization. Nat Rev Genet. 2001 May;2(5):333-41. Review. PubMed

5

Science of Proteomics: Major Dimensions & Trends

WHAT IS PROTEOMICS?

The dream of having genomes completely sequenced is now a reality. The complete sequence of many genomes including the human one is known. However, the understanding of probably half a million human proteins encoded by less than 30'000 genes is still a long way away and the hard work to unravel the complexity of biological systems is yet to come. A new fundamental concept called proteome (PROTEin complement to a genOME) has recently emerged that should drastically help to unravel biochemical and physiological mechanisms of complex multivariate diseases at the functional molecular level. The discipline of proteomics has been initiated to complement physical genomic research. Proteomics can be defined as the qualitative and quantitative comparison of proteomes under different conditions to further unravel biological processes.

"The analysis of complete complements of proteins. Proteomics includes not only the identification and quantification of proteins, but also the determination of their localization, modifications, interactions, activities, and, ultimately, their function. Initially encompassing just two- dimensional (2D) gel electrophoresis for protein separation and identification, proteomics now refers to any procedure that characterizes large sets of proteins. The explosive growth of this field is driven by multiple forces - genomics and its revelation of more and more new proteins; powerful protein technologies, such as newly developed mass spectrometry approaches, global [yeast] two- hybrid techniques, and spin- offs from DNA arrays; and innovative computational tools and methods to process, analyze, and interpret prodigious amounts of data."

The theme of molecular biology research, in the past, has been oriented around the gene rather than the protein. This is not to say that researchers have neglected to study proteins, but rather that the approaches and techniques most commonly used have looked primarily at the nucleic acids and then later at the protein(s) implicated.

The main reason for this has been that the technologies available, and the inherent characteristics of nucleic acids, have made the genes the low hanging fruit. This situation has changed recently and continues to change as larger scale, higher throughput methods are developed for both nucleic acids and proteins. The majority of processes that take place in a cell are not performed by the genes themselves, but rather by the proteins that they code for.

A disease can arise when a gene/protein is over- or under-expressed, or when a mutation in a gene results in a malformed protein, or when post translational modifications alter a protein's function. Thus to truly understand a biological process, the relevant proteins must be studied directly. But there are more challenges when studying proteins compared to studying genes, due to their complex 3-D structure which is related to the function, analogous to a machine.

Proteomics is defined as the systematic large-scale analysis of protein expression under normal and perturbed (stressed, diseased, and/or drugged) states, and generally involves the separation, identification, and characterization of all of the proteins in a cell or tissue sample. The meaning of the term has also been expanded, and is now used loosely to refer to the approach of analyzing which proteins a particular type of cell synthesizes, how much the cell synthesizes, how cells modify proteins after synthesis, and how all of those proteins interact.

There are orders of magnitude more proteins than genes in an organism - based on alternative splicing (several per gene) and post translational modifications (over 100 known), there are estimated to be a million or more.

Fortunately there are features such as folds and motifs, which allow them to be categorized into groups and families, making the task of studying them more tractable. There is a broad range of technologies used in proteomics, but the central paradigm has been the use of 2-D gel electrophoresis (2D-GE) followed by mass spectrometry (MS). 2D-GE is used to first separate the proteins by isoelectric point and then by size.

The individual proteins are subsequently removed from the gel and prepared, then analyzed by MS to determine their identity and characteristics. There are various types of mass analyzers used in proteomics MS including quadrupole, time-of-flight (TOF), and ion trap, and each has its own particular capabilities. Tandem arrangements are often used, such as quadrupole-TOF, to provide more analytical power. The recent development of soft ionization techniques, namely matrix-assisted laser desorption ionization (MALDI) and electro-spray ionization (ESI), has allowed large biomolecules to be introduced into the mass analyzer without completely decomposing their structures, or even without breaking them at all, depending on the design of the experiment.

There are techniques which incorporate liquid chromatography (LC) with MS, and others that use LC by itself. Robotics have been applied to automate several steps in the 2DGE-MS process such as spot excision and enzyme digests. To determine a protein's structure, XRD and NMR techniques are being improved to reach higher throughput and better performance.

For example, automated high-throughput crystallization methods are being used upstream of XRD to alleviate that bottleneck. For NMR, cryo-probes and flow probes shorten analysis time and decrease sample volume requirements. The hope is that determining about 10,000 protein structures will be enough to characterize the estimated 5,000 or so folds, which will feed into more reliable in silico structural prediction methods.

Structure by itself does not provide all of the desired information, but is a major step in the right direction. Protein chips are being developed for many of the processes in proteomics. For example, researchers are developing protocols for protein microarrays at institutions such as Harvard and Stanford as well as at several companies. These chips - grids of attached peptide fragments, attached antibodies, or gel "pads" with proteins suspended inside - will be used for various experiments such as protein-protein interaction studies and differential expression analysis.

They can also be used to filter out high abundance proteins before further experiments; one of the major challenges in proteomics is isolating and analyzing the low abundance proteins, which are thought to be the most important. There are many other types of protein chips, and the number will continue to grow. For example, microfluidics chips can combine the sample preparation steps prior to MS, such as enzyme digests, with nanoelectrospray ionization, all on the one chip. Or, the samples can be ionized directly off of the surface of the chip, similar to a MALDI target. Microfluidics chips are also being combined with NMR.

In the next few years, various protein chips will be used increasingly in diagnostic applications as well. The bioinformatics side of proteomics includes both databases and analysis software. There are many public and private databases containing protein data ranging from sequences, to functions, to post translational modifications. Typically, a researcher will first perform 2D-GE followed by MS; this will result in a fingerprint, molecular weight, or even sequence for each protein of interest, which can then be used to query databases for similarities or other information.

Swiss-Prot and TrEMBL, developed in a collaboration between the Swiss Institute of Bioinformatics and the European Bioinformatics Institute, are currently the major databases dedicated to cataloging protein data, but there are dozens of more specialized databases and tools. New bioinformatics approaches are constantly being introduced. Recent customized versions of PSI-BLAST can, for example, utilize not only the curated protein entries in Swiss-Prot but also linguistic analyses of biomedical journal articles to help determine protein family relationships. Publicly available databases and tools are popular, but there are also several companies offering subscriptions to proprietary databases, which often include protein-protein interaction maps generated using the yeast two-hybrid (Y2H) system.

The proteomics market is comprised of instrument manufacturers, bioinformatics companies, laboratory product suppliers, service providers, and other biotech related companies which can defy categorization. A given company can often overlap more than one of these areas. Many of the companies involved in the proteomics market are actually doing drug discovery as their major focus, while partnering, or providing services or subscriptions, to other companies to generate short term revenues. The market for proteomics products and services was estimated to be $1.0B in 2000, growing at a CAGR of 42% to about $5.8B in 2005.

The major drivers will continue to be the biopharmaceutical industry's pursuit of blockbuster drugs and the recent technological advances which have allowed large-scale studies of genes and proteins. Alliances are becoming increasingly important in this field, because it is challenging for companies to find all of the necessary expertise to cover the different activities involved in proteomics. Synergies must be created by combining forces. For example, many companies working with mass spectrometry, both the manufacturers and end user labs, are collaborating with protein chip related companies. The technologies are a natural fit for many applications, such as microfluidic chips which provide nanoelectrospray ionization into a mass spectrometer.

There are many combinations of diagnostics, instrumentation, chip, and bioinformatics companies which create effective partnerships. In general, proteomics appears to hold great promise in the pursuit of biological knowledge. There has been a general realization that the large-scale approach to biology, as opposed to the strictly hypothesis-driven approach, will rapidly generate much more useful information.

The two approaches are not mutually exclusive, and the happy medium seems to be the formation of broad hypotheses which are subsequently investigated by designing large-scale experiments and selecting the appropriate data. Proteomics and genomics, and other varieties of 'omics', will all continue to complement each other in providing the tools and information for this type of research.

Proteomics is the large-scale study of proteins, particularly their structures and functions. Proteins are vital parts of living organisms, as they are the main components of the physiological metabolic pathways of cells. The term "proteomics" was coined to make an analogy with genomics, the study of the genes. The word "proteome" is a portmanteau of "*prote*in" and "gen*ome*". The proteome of an organism is the set of proteins produced by it during its life, and its genome is its set of genes.

Proteomics is often considered the next step in the study of biological systems, after genomics. It is much more complicated than genomics, mostly because while an organism's genome is rather constant, a proteome differs from cell to cell and constantly changes through its biochemical interactions with the genome and the environment. One organism has radically different protein expression in different parts of its body, different stages of its life cycle and different environmental conditions. Another major difficulty is the complexity of proteins relative to nucleic acids. E.g., in human there are about 25 000 identified genes but an estimated >500 000 proteins that are derived from these genes. This increased complexity derives from mechanisms such as alternative splicing, protein modification (glycosylation,phosphorylation) and protein degradation.

Scientists are very interested in proteomics because it gives a much better understanding of an organism than genomics. First, the level of transcription of a gene gives only a rough estimate of its level of expression into a protein. An mRNA produced in abundance may be degraded rapidly or translated inefficiently, resulting in a small amount of protein. Second, many proteins experience post-translational modifications that profoundly affect their activities; for example some proteins are not active until they become phosphorylated. Methods such as phosphoproteomics and glycoproteomics are used to study post-translational modifications. Third, many transcripts give rise to more than one protein, through alternative splicing or alternative post-translational modifications. Finally, many proteins form complexes with other proteins or RNA molecules, and only function in the presence of these other molecules.

Since proteins play a central role in the life of an organism, proteomics is instrumental in discovery of biomarkers, such as markers that indicate a particular disease.

With the completion of a rough draft of the human genome, many researchers are looking at how genes and proteins interact to form other proteins. A surprising finding of the Human Genome Project is that there are far fewer protein-coding genes in the human genome than proteins in the human proteome (20,000 to 25,000 genes vs. > 500,000 proteins). The human body may even contain more than 2 million proteins, each having different functions. The protein diversity is thought to be due to alternative splicing and post-translational modification of proteins. The discrepancy implies that protein diversity cannot be fully characterized by gene expression analysis, thus proteomics is useful for characterizing cells and tissues.

To catalog all human proteins, their functions and interactions is a great challenge for scientists. An international collaboration with these goals is co-ordinated by the Human Proteome Organization (HUPO).

Studying Proteomics

Most proteins function in collaboration with other proteins, and one goal of proteomics is to identify which proteins interact. This often gives important clues about the functions of newly discovered proteins. Several methods are available to probe protein-protein interactions. The traditional method is yeast two-hybrid analysis. New methods include protein microarrays, immunoaffinity chromatography followed by mass spectrometry, and combinations of experimental methods such as phage display and computational methods.

Current research in proteomics requires first that proteins be resolved, sometimes on a massive scale. Protein separation can be performed using two-dimensional gel electrophoresis, which usually separates proteins first by isoelectric point and then by molecular weight. Protein spots in a gel can be visualized using a variety of chemical stains or fluorescent markers. Proteins can often be quantified by the intensity of their stain. Once proteins are separated and quantified, they are identified. Individual spots are cut out of the gel and cleaved into peptides with proteolytic enzymes. These peptides can then be identified by mass spectrometry, specifically matrix-assisted laser desorption-ionization time-of-flight (MALDI-TOF) mass spectrometry. In this procedure, a peptide is placed on a matrix, which causes the peptide to form crystals. Then the peptide on the matrix is ionized with a laser beam and an increase in voltage at the matrix is used to shoot the ions toward a detector in which the time it takes an ion to reach the detector depends on its mass. The higher the mass, the longer the time of flight of the ion. In a MALDI-TOF mass spectrometer, the ions can also be deflected with an electrostatic reflector that also focuses the ion beam. Thus, the masses of the ions reaching the second detector can be determined with high precision and these masses can reveal the exact chemical compositions of the peptides, and therefore their identities.

Protein mixtures can also be analyzed without prior separation. These procedures begin with proteolytic digestion of the proteins in a complex mixture. The resulting peptides are often injected onto a high pressure liquid chromatography column

(HPLC) that separates peptides based on hydrophobicity. HPLC can be coupled directly to a time-of-flight mass spectrometer using electrospray ionization. Peptides eluting from the column can be identified by tandem mass spectrometry (MS/MS). The first stage of tandem MS/MS isolates individual peptide ions, and the second breaks the peptides into fragments and uses the fragmentation pattern to determine their amino acid sequences. Labeling with isotope tags can be used to quantitatively compare proteins concentration among two or more protein samples.

One of the most promising developments to come from the study of human genes and proteins has been the identification of potential new drugs for the treatment of disease. This relies on genome and proteome information to identify proteins associated with a disease, which computer software can then use as targets for new drugs. For example, if a certain protein is implicated in a disease, its 3D structure provides the information to design drugs to interfere with the action of the protein. A molecule that fits the active site of an enzyme, but cannot be released by the enzyme, will inactivate the enzyme. This is the basis of new drug-discovery tools, which aim to find new drugs to inactivate proteins involved in disease. As genetic differences among individuals are found, researchers expect to use these techniques to develop personalized drugs that are more effective for the individual.

A computer technique which attempts to fit millions of small molecules to the three-dimensional structure of a protein is called "virtual ligand screening". The computer rates the quality of the fit to various sites in the protein, with the goal of either enhancing or disabling the function of the protein, depending on its function in the cell. A good example of this is the identification of new drugs to target and inactivate the HIV-1 protease. The HIV-1 protease is an enzyme that cleaves a very large HIV protein into smaller, functional proteins. The virus cannot survive without this enzyme; therefore, it is one of the most effective protein targets for killing HIV.

There are many distributed computing programs, such as the world community grid, which allows people around the world to help scientists by computing calculations. The software adds to the use of super computers by using the unused processing power of millions of home computers. The world community grid works on HIV, cancer, and protein folding. All three projects centre around protein modelling and protein modification models. Using the data gained from distributed computing models of proteins, scientists can develop more specific and effective therapies. In addition, most enzymes act as part of complexes and networks, which also affect the way an enzyme acts in a cell. Understanding these complex networks will assist in developing drugs that affect the function of these complexes.

Biomarkers

Understanding the proteome, the structure and function of each protein and the complexities of protein-protein interactions will be critical for developing the most effective diagnostic techniques and disease treatments in the future.

An interesting use of proteomics is using specific protein biomarkers to diagnose disease. A number of techniques allow to test for proteins produced during a particular disease, which helps to diagnose the disease quickly. Techniques include western blot, immunohistochemical staining, enzyme linked immunosorbent assay (ELISA) or mass spectrometry. The following are some of the diseases that have characteristic biomarkers that physicians can use for diagnosis:

- In Alzheimer's disease, elevations in beta secretase creates amyloid/beta-protein, which causes plaque to build up in the patient's brain, which causes dementia. Targeting this enzyme decreases the amyloid/beta-protein and so slows the progression of the disease. A procedure to test for the increase in amyloid/beta-protein is immunohistochemical staining, in which antibodies bind to specific antigens or biological tissue of amyloid/beta-protein.
- Heart disease is commonly assessed using several key protein based biomarkers. Standard protein biomarkers for CVD include interleukin-6, interleukin-8, serum amyloid A protein, fibrinogen, and troponins. cTnI cardiac troponin I increases in concentration within 3 to 12 hours of initial cardiac injury and can be found elevated days after an acute myocardial infarction. A number of commercial antibody based assays as well as other methods are used in hospitals as primary tests for acute MI.
- Proteomic analysis of kidney cells and cancerous kidney cells is producing promising leads for biomarkers for renal cell carcinoma and developing assays to test for this disease. In kidney-related diseases, urine is a potential source for such biomarkers. Recently, it has been shown that the identification of urinary polypeptides as biomarkers of kidney-related diseases allows to diagnose the severity of the disease several months before the appearance of the pathology.Article

Branches

1. *Protein separation*. Proteomic technologies rely on the ability to separate a complex mixture so that individual proteins are more easily processed with other techniques.
2. *Protein identification*. Well-known methods include low-throughput sequencing through Edman degradation. Higher-throughput proteomic techniques are based on mass spectrometry, commonly peptide mass fingerprinting on MALDI-TOF instruments, or De novo repeat detection MS/MS on instruments capable of more than one round of mass spectrometry. MS/MS data can be

analyzed by simple database searches as is the case for PMFs and additionally they can be analyzed by de novo sequencing and homology searching. This particular approach allows to even identify similar (homolog) proteins, e.g. across species in case a protein was derived from an organism with unsequenced genome. Antibody-based assays can also be used, but are unique to one sequence motif.

3. In quantitative proteomics different methods are used to obtain quantitative information on a proteome-wide scale. Rather than just lists of proteins, quantitative proteomics provides functional information and reveals temporal changes in the proteome.
4. *Protein sequence analysis* is a branch of bioinformatics that deals with searching databases for possible protein or peptide matches by algorithms such as Mascot, PEAKS(software), OMSSA, SEQUEST and X!Tandem, PWB Protein Identification Cluster Software Solution, functional assignment of domains, prediction of function from sequence, and evolutionary relationships of proteins.
5. *Structural proteomics* concerns the high-throughput determination of protein structures in three-dimensional space. Common methods are x-ray crystallography and NMR spectroscopy.
6. *Interaction proteomics* concerns the investigation of protein interactions on the atomic, molecular and cellular levels. see related article on Protein-protein interaction prediction.
7. *Protein modification* studies the modified forms of proteins. Almost all proteins are modified from their pure translated amino-acid sequence, by so-called post-translational modification. Specialized methods have been developed to study phosporylation (phosphoproteomics) and glycosylation (glycoproteomics).
8. *Cellular proteomics* is a new branch of proteomics aiming to map the location of proteins and protein-protein interactions in whole cells during key cell events. Centers around the use of techniques such as X-ray Tomography and optical fluorescence microscopy.
9. *Experimental bioinformatics* is a branch of bioinformatics, as it is applied in proteomics, coined by Mathias Mann. It involves the mutual design of experimental and bioinformatics methods to create (extract) new types of information from proteomics experiments.

Technologies

Proteomics uses various technologies:

- One- and two-dimensional gel electrophoresis is used to identify the relative mass of a protein and its isoelectric point.
- X-ray crystallography and nuclear magnetic resonance are used to characterize the three-dimensional structure of peptides and proteins. However, low-resolution techniques such as circular dichroism, Fourier transform infrared spectroscopy and Small angle X-ray scattering (SAXS) can be used to study the secondary structure of proteins.
- Tandem mass spectrometry combined with reverse phase chromatography or 2-D electrophoresis is used to identify proteins using database search tools such as Mascot, Phenyx, PEAKS(software), OMSSA, X!Tandem and SEQUEST or de novo algorithms and quantify all the levels of proteins found in cells.
- Scaffold, a software tool useful in visualization of tandem mass spectrometry results.
- Tandem mass spectrometry combined with tagging technologies such as TMT, ICPL or iTRAQ is used for quantification of proteins and peptides.
- Mass spectrometry (no-tandem), often MALDI-TOF, is used to identify proteins by peptide mass fingerprinting. This technology is also used in so-called "MALDI-TOF MS protein profiling" where samples (i.e. serum) are prepared by either protein chips (SELDI-TOF MS), magnetic beads (The Bruker Daltonics protein profiling platform) or with other methods of sample treatment, such as liquid chromatography, size-exclusion and immunoaffinity. Protein peaks of interest must be identified by tandem mass spectrometry. Protein profiling with MALDI-TOF MS could be of high use in clinical diagnostics, but so far there has been little success with advancing MALDI-TOF MS protein profiling into clinical validation due to high analytical variation.
- ICP-MS combined with MeCAT - Metal Coded Tagging - technology is used for ultrasensitive quantification of proteins and peptides down to low attomol range
- Affinity chromatography, yeast two hybrid techniques, fluorescence resonance energy transfer (FRET), and Surface Plasmon Resonance (SPR) are used to identify protein-protein and protein-DNA binding reactions.
- X-ray Tomography used to determine the location of labeled proteins or protein complexes in an intact cell. Frequently correlated with images of cells from light based microscopes.
- Software based image analysis is utilized to automate the quantification and detection of spots within and among gels samples. While this technology

is widely utilized, the intelligence has not been perfected yet. For example, the leading software tools in this area tend to agree on the analysis of well-defined, well-separated protein spots, but they deliver different results and tendencies with less-defined less-separated spots - thus necessitating manual verification of results.

SHOTGUN PROTEOMICS

Shotgun proteomics is a method of identifying proteins in complex mixtures using a combination of high performance liquid chromatography combined with mass spectrometry. The name is derived from shotgun sequencing of DNA which is itself named by analogy with the rapidly-expanding, quasi-random firing pattern of a shotgun. In shotgun proteomics, the proteins in the mixture are digested and the resulting peptides are separated by liquid chromatography. Tandem mass spectrometry is then used to identify the peptides.

PHOSPHOPROTEOMICS

Phosphoproteomics is a branch of proteomics that identifies, catalogs, and characterizes proteins containing a phosphate group as a post-translational modification. Phosphorylation is a key reversible modification that regulates protein function, subcellular localization, complex formation, degradation of proteins and therefore cell signalling networks. With all of these modification results, it is assumed that up to 30% of all proteins may be phosphorylated, some multiple times.

Compared to expression analysis, phoshoproteomics provides two additional layers of information. First, it provides clues on what protein or pathway might be activated because a change in phosphorylation status almost always reflects a change in protein activity. Second, it indicates what proteins might be potential drug targets as exemplified by the kinase inhibitor Gleevec. While phosphoproteomics will greatly expand knowledge about the numbers and types of phosphoproteins, its greatest promise is the rapid analysis of entire phosphorylation based signalling networks.

Overview of Phosphoproteomic Analysis

A sample large-scale phosphoproteomic analysis includes

1. Cultured cells undergo SILAC encoding.
2. Cells are stimulated with factor of interest (eg. growth factor, hormone).
3. Stimulation can occur for various lengths of time for temporal analysis.

4. Cells are lysed and enzymatically digested.
5. Peptides are separated using ion exchange chromatography.
6. Phosphopeptides are enriched using phosphospecific antibodies, immobilized metal affinity chromatography or titanium dioxide (TiO_2) chromatography.
7. Phosphopeptides are analyzed using mass spectrometry.
8. Peptides are sequenced and analyzed.

Tools and Methods

The analysis of the entire complement of phosphorylated proteins in a cell is certainly a feasible option. This is due to the optimization of enrichment protocols for phosphoproteins and phosphopeptides, better fractionation techniques using chromatography, and improvement of methods to selectively visualize phosphorylated residues using mass spectrometry. Although the current procedures for phosphoproteomic analysis are greatly improved, there is still sample loss and inconsistencies with regards to sample preparation, enrichment , and instrumentation. Bioinformatics tools and biological sequence databases are also necessary for high-throughput phosphoproteomic studies.

Enrichment Strategies

Previous procedures to isolate phosphorylated proteins included radioactive labeling with P-labeled ATP followed by SDS polyacrylamide gel electrophoresis or thin layer chromatography. These traditional methods are inefficient because it is impossible to obtain large amounts of proteins required for phosphorylation analysis. Therefore, the current and simplest methods to enrich phosphoproteins are affinity purification using phosphospecific antibodies, immobilized metal affinity chromatography (IMAC), strong cation exchange (SCX) chromatography, or titanium dioxide chromatography. Antiphosphotyrosine antibodies have been proven very successful in purification, but fewer reports have been published using antibodies against phosphoserine- or phosphothreonine-containing proteins. IMAC enrichment is based on phosphate affinity for immobilized metal chelated to the resin. SCX separates phosphorylated from non-phosphorylated peptides based on the negatively charged phosphate group. Titanium dioxide chromatography is a newer technique that requires significantly less column preparation time. Many phosphoproteomic studies use a combination of these enrichment strategies to obtain the purest sample possible.

Mass Spectrometry Analysis

Mass spectrometry is currently the best method to adequately compare pairs of protein samples. The two main procedures to perform this task are using isotope-coded affinity tags (ICAT) and stable isotopic amino acids in cell culture (SILAC). In the ICAT procedure samples are labeled individually after isolation with mass-coded reagents that modify cysteine residues. In SILAC, cells are cultured separately in the presence of different isotopically labeled amino acids for several cell divisions allowing cellular proteins to incorporate the label. Mass spectrometry is subsequently used to identify phosphoserine, phosphothreonine, and phosphotyrosine-containing peptides.

Phosphoproteomics in the Study of Signal Transduction

Intracellular signal transduction is primarily mediated by the reversible phosphorylation of various signalling molecules by enzymes dubbed kinases. Kinases transfer phosphate groups from ATP to specific serine, threonine or tyrosine residues of target molecules. The resultant phosphorylated protein may have altered activity level, subcellular localization or tertiary structure.

Phosphoproteomic analyses are ideal for the study of the dynamics of signalling networks. In one study design, cells are exposed to SILAC labelling and then stimulated by a specific growth factor. The cells are collected at various timepoints, and the lysates are combined for analysis by tandem MS. This allows experimenters to track the phosphorylation state of many phosphoproteins in the cell over time. The ability to measure the global phosphorylation state of many proteins at various time points makes this approach much more powerful than traditional biochemical methods for analyzing signalling network behavior.

One study was able to simultaneously measure the fold-change in phosphorylation state of 127 proteins between unstimulated and EphrinB1-stimulated cells. Of these 127 proteins, 40 showed increased phosphorylation with stimulation by EphrinB1. The researchers were able to use this information in combination with previously published data to construct a signal transduction network for the proteins downstream of the EphB2 receptor.

Another recent phosphoproteomic study included large-scale identification and quantification of phosphorylation events triggered by the anti-diuretic hormone vasopressin in kidney collecting duct. A total of 714 phosphorylation sites on 223 unique phosphoproteins were identified, including three novel phosphorylation sites in the vasopressin-sensitive water channel aquaporin-2 (AQP2).

Phosphoproteomics in the Study of Cancer

Since the inception of phosphoproteomics, cancer research has focused on changes to the phosphoproteome during tumor development. Phosphoproteins could be cancer markers useful to cancer diagnostics and therapeutics. In fact, research has shown that there are distinct phosphotyrosine proteomes of breast and liver tumors. There is also evidence of hyperphosphorylation at tyrosine residues in breast tumors but not in normal tissues. Findings like these suggest that it is possible to mine the tumor phosphoproteome for potential biomarkers.

Increasing amounts of data are available suggesting that distinctive phosphoproteins exist in various tumors and that phosphorylation profiling could be used to fingerprint cancers from different origins. In addition, systematic cataloguing of tumor-specific phosphoproteins in individual patients could reveal multiple causative players during cancer formation. By correlating this experimental data to clinical data such as drug response and disease outcome, potential cancer markers could be identified for diagnosis, prognosis, prediction of drug response, and potential drug targets.

Limitations

While phosphoproteomics has greatly expanded knowledge about the numbers and types of phosphoproteins, along with their role in signaling networks, there are still a few limitations to these techniques. To begin with, isolation methods such as anti-phosphotyrosine antibodies do not distinguish between isolating tyrosine-phosphorylated proteins and proteins associated with tyrosine-phosphorylated proteins. Therefore, even though phosphorylation dependent protein-protein interactions are very important, it is important to remember that a protein detected by this method is not necessarily a direct substrate of any tyrosine kinase. Only by digesting the samples before immunoprecipitation can isolation of only phosphoproteins and temporal profiles of individual phosphorylaiton sites be produced. Another limitation is that some relevant proteins will likely be missed since no extraction condition is all encompassing. It is possible that proteins with low stoichiometry of phosphorylation, in very low abundance, or phosphorylated as a target for rapid degradation will be lost.

BOTTOM-UP PROTEOMICS

Bottom-up proteomics is a method for the identification of proteins in complex mixtures through a combination of two-dimensional gel electrophoresis and mass spectrometry. The name is derived from the analogous bottom-up DNA sequencing. Mass spectrometry analysis consists of a single stage of mass spectrometry for peptide mass fingerprinting or tandem mass spectrometry to obtain a peptide sequence tag.

TOP-DOWN PROTEOMICS

Top-down proteomics is a method of protein identification that uses an ion trapping mass spectrometer to store an isolated protein ion for mass measurement and tandem mass spectrometry analysis. The name is derived from the similar approach to DNA seqencing. Proteins are typically ionized by electrospray ionization and trapped in a Fourier transform ion cyclotron resonance (Penning trap) or quadrupole ion trap (Paul trap) mass spectrometer. Fragmentation for tandem mass spectrometry is accomplished by electron capture dissociation or electron transfer dissociation.

CYTOMICS

Cytomics is the study of cell systems (cytomes) at a single cell level. It combines all the bioinformatic knowledge to attempt to understand the molecular architecture and functionality of the cell system (Cytome). Much of this is achieved by using molecular and microscopic techniques that allow the various components of a cell to be visualised as they interact *in vivo*.

TRANSCRIPTOME

The transcriptome is the set of all messenger RNA (mRNA) molecules, or "transcripts", produced in one or a population of cells. The term can be applied to the total set of transcripts in a given organism, or to the specific subset of transcripts present in a particular cell type. Unlike the genome, which is roughly fixed for a given cell line (excluding mutations), the transcriptome can vary with external environmental conditions. Because it includes all *mRNA* transcripts in the cell, the transcriptome reflects the genes that are being actively expressed at any given time, with the exception of mRNA degradation phenomena such as transcriptional attenuation. The study of *transcriptomics* examines the expression level of mRNAs in a given cell population, often using high-throughput techniques based on DNA microarray technology.

Applications and Analysis

The transcriptomes of stem cells and cancer cells are of particular interest to researchers who seek to understand the processes of cellular differentiation and carcinogenesis. A number of organism-specific transcriptome databases have been constructed and annotated to aid in the identification of genes that are differentially expressed in distinct cell populations or subtypes; however, the analysis of relative mRNA expression levels can be complicated by the fact that relatively small changes in mRNA expression can produce large changes in the total amount of the corresponding protein present in the cell. One analysis method, known as Gene Set Enrichment Analysis, identifies coregulated gene networks rather than individual genes that are up- or down-regulated in different cell populations.

mRNA Regulation

Although microarray studies can reveal the relative amounts of different mRNAs in the cell, levels of mRNA are not directly proportional to the expression level of the proteins they code for. The number of protein molecules synthesized using a given mRNA molecule as a template is highly dependent on translation-initiation features of the mRNA sequence; in particular, the ability of the translation initiation sequence is a key determinant in the recruiting of ribosomes for protein translation. The complete protein complement of a cell or organism is known as the proteome.

A study of 158,807 mouse transcripts revealed that 4520 of these transcripts form antisense partners that are base pair complementary to the exons of genes. These results raise the possibility that significant numbers of "antisense RNA-coding genes" might participate in the regulation of the levels of expression of protein-coding mRNAs.

PEGYLATION

History

In 1970s, pioneering research by Davis, Abuchowski and colleagues foresaw the potential of the conjugation of Polyethylene glycol (PEG) to Proteins. This technique is now well established and is known as PEGylation.

PEGylation, is a process of attaching the strands of the polymer PEG to molecules most typically Peptides, Proteins, and Antibody fragments, that can help to meet the challenges of improving the safety and efficiency of many therapeutics. It produces alterations in the physiochemical properties including changes in conformation, electrostatic binding, hydrophobicity etc. These physical and chemical changes increase systemic retention of the therapeutic agent. Also, it can influence the binding affinity of the therapeutic moiety to the cell receptors and can alter the absorption and distribution patterns.

PEGylation, by increasing the molecular weight of a molecule, can impart several significant pharmacological advantages over the unmodified form, such as:

- Improved drug solubility
- Reduced dosage frequency, without diminished efficacy with potentially reduced toxicity
- Extended circulating life
- Increased drug stability
- Enhanced protection from proteolytic degradation

The PEGylated drugs are having the following commercial advantages also:

- Opportunities for new delivery formats and dosing regimens
- Extended patent life of previously approved drugs

PEGylated Pharmaceuticals on the Market

The clinical value of PEGylation is now well established. ADAGEN (PEG- bovine adenosine deaminase) manufactured by Enzon Pharmaceuticals, Inc., US was the first PEGylated protein approved by FDA in March 1990, to enter the market. It is used to treat X-linked severe combined immunogenicity syndrome, as an alternative to bone marrow transplantation and enzyme replacement by gene therapy. Since the introduction of ADAGEN, a large number of PEGylated protein and peptide pharmaceuticals have followed and many others are under clinical trial or under development stages. Some of the successful examples are:

- PEGASYS: PEGylated alpha-interferons for use in the treatment of hepatis C (Hoffman-La Roche)
- PEG-Intron: PEGylated alpha -interferons for chronic hepatis C (Schering-Plough / Enzon)
- Oncaspar: PEGylated L-asparaginase for the treatment of acute lymphoblastic leukemia in patients who are hypersensitive to the native unmodified form of L-asparaginase (Enzon)
- Neulasta: PEGylated recombinant methionyl human granulocyte colony stimulating factor for severe cancer chemotherapy induced neutropenia (Amgen)
- Doxil: PEGylated liposome containing doxorubicin for the treatment of Cancer (Sequus)

PEG Moiety Properties

PEG is a particularly attractive polymer for conjugation. The specific characteristics of PEG moieties relevant to pharmaceutical applications are:

- Water solubility
- High mobility in solution
- Lack of toxicity and immunogenicity
- Ready clearance from the body
- Altered distribution in the body

PEGylation Process

The first step of the PEGylation is the suitable functionalization of the PEG polymer at one or both terminals. PEGs that are activated at each terminus with the same reactive moiety is known as "homobifunctional", where as if the functional groups present are different, then the PEG derivative is referred as "heterobifunctional" or "heterofunctional." The chemically active or activated derivatives of the PEG polymer are prepared to attach the PEG to the desired molecule.

The choice of the suitable functional group for the PEG derivative is based on the type of available reactive group on the molecule that will be coupled to the PEG. For proteins, typical reactive amino acids include lysine, cysteine, histidine, arginine, aspartic acid, glutamic acid, serine, threonine, tyrosine. The N-terminal amino group and the C-terminal carboxylic acid can also be used.

The techniques used to form first generation PEG derivatives are generally reacting the PEG polymer with a group that is reactive with hydroxyl groups, typically anhydrides, acid chlorides, chloroformates and carbonates. In the second generation PEGylation chemistry more efficient functional groups such as aldehyde, esters, amides etc made available for conjugation.

As applications of PEGylation have become more and more advanced and sophisticated, there has been an increase in need for heterobifunctional PEGs for conjugation. These heterobifunctional PEGs are very much useful in linking two entities, where a hydrophilic, flexible and biocompatible spacer is needed. Preferred end groups for heterobifunctional PEGs are maleimide, vinyl sulphones, pyridyl disulphide, amine, carboxylic acids and NHS esters.

LIST OF PROTEINS

A list of proteins (and protein complexes). This list aims to organize information on the protein universe. All proteins can be found in the human proteome unless marked with a "%".

If a protein has an EC number, it should be on List of enzymes and not on this page, even if it fits into one of the categories below.

Fibrous Protein

Cytoskeletal Proteins

- tubulin
- actin

- FtsZ %
- keratin
- myosin
- Tau (protein)
- dystrophin

Extracellular Matrix Proteins

- Collagen
- Elastin
- Reelin
- F-spondin

Globular Proteins

- Serum albumin
- Serum Amyloid P Component

Coagulation Factors

- Fibrin
- Thrombin
- Factor XIII
- Factor VIII
- von Willebrand Factor
- protein C
- Protein Z
- Protein Z-related protease inhibitor
- Protein S
- Complement proteins
    - C1-inhibitor
    - C3-convertase

Acute Phase Proteins

- C-reactive protein

Hemoproteins

- hemoglobin
- myoglobin

Cell Adhesion

- cadherin
- integrin
- NCAM
- selectin

Transmembrane Transport Proteins

Ion pumping enzymes are in the enzymes section.

- CFTR
- Glycophorin C
- Scramblase

Ion Channels

- Acetylcholine receptor
    - Nicotinic acetylcholine receptor
    - Muscarinic acetylcholine receptoris NOT an ion channel, but a G-protein coupled receptor; see below]
- Potassium channel

Synport/Antiport Proteins

- glucose transporter

Hormones and Growth Factors

- oxytocin
- insulin
- Epidermal growth factor
- Insulin-like growth factor

Transmembrane Receptors

- G-protein-coupled receptor
    - Rhodopsin

Intracellular Receptors

- estrogen receptor

DNA-binding Protein

- Histones

transcription Regulation

- CI protein %

Transcription regulatory proteins that are receptors are in the receptors section.

- C-myc
- MyoD
- FOXP2
- FOXP3
- P53

Immune System Proteins

- immunoglobins
- T cell receptor
- Major histocompatibility antigens

Nutrient Storage/Transport

- ferritin

Chaperone Proteins

- GroEL %

Complexes with Multiple Components Including Proteins

- spliceosome
- ribonucleoprotein (generic)
- Signal recognition particle
- nucleosome

HUMAN PROTEIN REFERENCE DATABASE

The Human Protein Reference Database is a protein database accessible through internet. This resource depicts information on human protein functions including protein-protein interactions, post-translational modifications, enzyme-substrate relationships and disease associations. Protein annotation information that is catalogued was derived through manual curation using published literature by expert biologists and through bioinformatics analyses of the protein sequence. HPRDalso integrates data from Human Proteinpedia, a community portal for integrating human protein data. The data from HPRD can be freely obtained and commercial entities may not use this site without prior licensing authorization.

REFERENCES

Abuchowski, McCoy, Palczuk, van Es and Davis (1977). "Effect of covalent attachment of polyethylene glycol on immunogenicity and circulating life of bovine liver catalase." Journal of Biological Chemistry 252(11): 3582-3586.

Arora, Pankaj S., et al. (2005). "Comparative evaluation of two two-dimensional gel electrophoresis image analysis software applications using synovial fluids from patients with joint disease". Journal of Orthopaedic Science 10 (2): 160-166.

Belhajjame, K. et al. Proteome Data Integration: Characteristics and Challenges. Proceedings of the UK e-Science All Hands Meeting, ISBN 1-904425-53-4, September 2005, Nottingham, UK.

Bernas T, Gregori G, Asem EK, Robinson JP., Integrating cytomics and proteomics, Mol Cell Proteomics. 2006 Jan;5(1):2-13.

Campbell and Reece. Biology. Sixth Edition. Pearson Education Inc. 2002. pg 392-393.

Chen X, Velliste M, Murphy RF., Automated interpretation of subcellular patterns in fluorescence microscope images for location proteomics, Cytometry A. 2006 Jul;69(7):631-40.

Davies E, Stankovic B, Azama K, Shibata K, Abe S., Novel components of the plant cytoskeleton: A beginning to plant "cytomics", Plant Science, Invited Review, Plant Science 2001; (160)2: 185-196.

Decramer S et al "Predicting the clinical outcome of congenital unilateral ureteropelvic junction obstruction in newborn by urinary proteome analysis" Nature Medicine 2006; 12:398-400 Article

Fee (2003). "Size-exclusion reaction chromatography (SERC): A new technique for protein PEGylation." Biotechnology and Bioengineering 82(2): 200-206.

Fee and Alstine (2006). "PEG-proteins: Reaction engineering and separation issues." Chemical Engineering Science 61(3): 924-939.

Herrera G, Diaz L, Martinez-Romero A, Gomes A, Villamon E, Callaghan RC, O'connor JE., Cytomics: A multiparametric, dynamic approach to cell research, Toxicol In Vitro. 2006 Jul 22.

Hoffert JD, et al. Quantitative phosphoproteomics of vasopressin-sensitive renal cells: regulation of aquaporin-2 phosphorylation at two sites. Proc Natl Acad Sci U S A. 2006 May 2;103(18):7159-64.

Hye A, Lynham S, Thambisetty M, et al. " Proteome-based plasma biomarkers for Alzheimer's disease." Brain 129: 3042-3050, (2006).

Introduction to Antibodies - Enzyme-Linked Immunosorbent Assay (ELISA). http://www.chemicon.com/resource/ANT101/a2C.asp. (Retrieved 29 Nov 2006)

Johnson, S & Hunter, T. (2004) Phosphoproteomics finds its timing. *Nature Biotech* 22(9): 1093-1094.

Kalume, D. et al. (2003) Tackling the phosphoproteome: tools and strategies *Current Opinion in Chemical Biology* 7: 64-69.

Kodera, Matsushima, Hiroto, Nishimura, Ishii, Ueno and Inada (1998). "Pegylation of proteins and bioactive substances for medical and technical applications." Progress in Polymer Science 23(7): 1233-1271.

Kriete A., Cytomics in the realm of systems biology, Cytometry A. 2005 Nov;68(1):19-20.

Liebler, D. C. 2002. Introduction to proteomics: tools for the new biology. Humana Press, Totowa, NJ. ISBN 0-585-41879-9 (electronic, on Netlibrary?), ISBN 0-89603-991-9 hardback,

Lim, Y. (2005) Mining the tumor phosphoproteome for cancer markers. *Clin Cancer Res* 11(9): 3163-3169.

Lim, Y. (2005) Mining the tumor phosphoproteome for cancer markers. *Clin Cancer Res* 11(9): 3163-3169

Macaulay IC, Carr P, Gusnanto A, et al. "Platelet Genomics and Proteomics in Human Health and Disease." The Journal of Clinical Investigation 115: 3370-3377, (December 2005).

Morar, Jeffrey and Mark (2006). "PEGylation of Proteins: A Structural Approach." Biopharm International 19(4): 34.

Mumby, M. & Brekken, D. (2005) Phosphoproteomics: new insights into cellular signalling. *Genome Biology* 6: 230.1-230.7.

Murphy RF., Cytomics and location proteomics: automated interpretation of subcellular patterns in fluorescence microscope images, Cytometry A. 2005 Sep;67(1):1-3.

Olsen JV et al. Global, in vivo, and site-specific phosphorylation dynamics in signaling networks. Cell. 2006 Nov 3;127(3):635-48.

Olsen, JV et al. (2006) Global, in vivo, and site-specific phosphorylation dynamics in signaling networks. *Cell* 127(3): 635-48.

Perroud B, Lee J, Valkova N, et al. "Pathway Analysis of Kidney Cancer Using Proteomics and *Metabolic Profiling." Biomed Central: 65-82, (24 November 2006).

Rediscovering Biology Online Textbook. Unit 2 Proteins and Proteomics. 1997-2006.

Roberts, Bentley and Harris (2002). "Chemistry for peptide and protein PEGylation." Advanced Drug Delivery Reviews 54(4): 459-476.

Rogers MA, Clarke P, Noble J, et al. "Proteomic Profiling of Urinary Proteins in Renal Cancer by Surface Enhanced Laser Desorption Ionization, and Neural-Network Analysis: Identification of Key Issues Affecting Clinical Potential Utility." Cancer Research 63: 6971-6983, (15 October 2003).

Schmelzle, K. & White, F. (2006) Phosphoproteomic approaches to ellucidate cellular signaling networks. *Current Opinion in Chemical Biology* 17: 406-414.

Schubert W., Cytomics in characterizing toponomes: towards the biological code of the cell, Cytometry A. 2006 Apr;69(4):209-11.

Twyman, R. M. 2004. Principles of proteomics. BIOS Scientific Publishers, New York. ISBN 1-85996-273-4.(covers almost all branches of proteomics)

Valet G, Predictive medicine by cytomics and the challenges of a human cytome project, Business Briefing: Future Drug Discovery 2004, Ed: E Cooper, World Markets Research Center Ltd, London, 2004; 46-51.

Valet G, Tarnok A, Cytomics in predictive medicine, Cytometry, 2003, 53B: 1-3.

Valet G., Cytomics: an entry to biomedical cell systems biology, Cytometry A. 2005 Feb; 63(2):67-8.

Van Osta P, Ver Donck K, Bols L, Geysen J., Cytomics and drug discovery., Cytometry A. 2006 Mar;69(3):117-8.

Vasan RS. "Biomarkers of cardiovascular disease: molecular basis and practical considerations" Circulation. 2006;113:2335-2362.

Veronese (2001). "Peptide and protein PEGylation: a review of problems and solutions." Biomaterials 22(5): 405-417.

Veronese and Harris (2002). "Introduction and overview of peptide and protein pegylation." Advanced Drug Delivery Reviews 54(4): 453-456.

Veronese and Pasut (2005). "PEGylation, successful approach to drug delivery." Drug Discovery Today 10(21): 1451-1458.

Weaver. R.F. Molecular Biology. Third Edition. The McGraw-Hill Companies Inc. 2005. pgs 840-849.

Westermeier, R. and T. Naven. 2002. Proteomics in practice: a laboratory manual of proteome analysis. Wiley-VCH, Weinheim. ISBN 3-527-30354-5.(focused on 2D-gels, good on detail)

Wiacek C, Muller S, Benndorf D, A cytomic approach reveals population heterogeneity of Cupriavidus necator in response to harmful phenol concentrations, Proteomics 2006 Nov;6(22):5983-94.

Wilkins MR, Williams KL, Appel RD, Hochstrasser DF. *Proteome research: new frontiers in functional genomics*. Berlin Heidelberg, Springer Verlag; 1997.

World Community Grid. http://www.worldcommunitygrid.org (Retrieved 29 Nov 2006)

Glossary

Achondroplasia: The most common and well known form of short limbed dwarfism characterized by a normal trunk size with disproportionally short arms and legs, and a disproportionally large head; autosomal dominant condition.

Advanced maternal age: Women over age 34 (age 35 at delivery) at increased risk for nondisjunction trisomy in fetus.

Alcoholism: A chronic and progressive condition characterized by the inability to control the consumption of alcohol.

Allele: An alternative form of a gene; any one of several mutational forms of a gene.

Alpha-fetoprotein (AFP): A protein excreted by the fetus into the amniotic fluid and from there into the mother's bloodstream through the placenta.

Alu repetitive sequence: The most common dispersed repeated DNA sequence in the human genome accounting for 5% of human DNA. The name is derived from the fact that these sequences are cleaved by the restriction endonuclease Alu.

Amino acid sequence: The linear order of the amino acids in a protein or peptide.

Amniocentesis: Prenatal diagnosis method using cells in the amniotic fluid to determine the number and kind of chromosomes of the fetus and, when indicated, perform biochemical studies.

Amniocyte: Cells obtained by amniocentesis.

Amplification: Any process by which specific DNA sequences are replicated disproportionately greater than their representation in the parent molecules.

Aneuploidy: State of having variant chromosome number (too many or too few). (i.e. Down syndrome, Turner syndrome).

Angelman syndrome: A condition characterized by severe mental deficiency, developmental delay and growth deficiency, puppet-like gait and frequent laughter unconnected to emotions of happiness.

Apert syndrome: A condition caused by the premature closure of the sutures of the skull bones, resulting in an altered head shape, with webbed fingers and toes. Autosomal dominant.

Artificial insemination: The placement of sperm into a female reproductive tract or the mixing of male and female gametes by other than natural means.

Autoradiograph: A photographic picture showing the position of radioactive substances in tissues.

Autosome: A nuclear chromosome other than the X- and Y-chromosomes.

Bacteriophage: A virus whose host is a bacterium; commonly called phage.

Barr body: The condensed single X-chromosome seen in the nuclei of somatic cells of female mammals. base pair a pair of hydrogen-bonded nitrogenous bases (one purine and one pyrimidine) that join the component strands of the DNA double helix.

Base sequence: A partnership of organic bases found in DNA and RNA; adenine forms a base pair with thymine (or uracil) and guanine with cytosine in a double-stranded nucleic acid molecule.

Baysian analysis: A mathematical method to further refine recurrence risk taking into account other known factors.

Becker muscular dystrophy: X-linked condition characterized by progressive muscle weakness and wasting; manifests later in life with progression less severe than Duchenne muscular dystrophy.

Biological atlas: Maps describing different aspects of protein function should be compiled into a "biological atlas" By integrating the information contained in the atlas, increasingly meaningful biological hypotheses could be formulated.

Carrier: An individual heterozygous for a single recessive gene.

cDNA maps: Shows the locations of expressed DNA regions (exons) on the chromosomal map. Because they represent expressed genomic regions, cDNAs are thought to identify the parts of the genome with the most biological and medical significance. A cDNA map can provide the chromosomal location for genes whose functions are currently unknown. For disease- gene hunters, the map can also suggest a set of candidate genes to test when the approximate location of a disease gene has been mapped by genetic linkage techniques.

cDNA: Complementary DNA produced from a RNA template by the action of RNA-dependent DNA polymerase.

Cell mapping: The determination of the subcellular location of proteins and of protein- protein interactions by the purification of organelles or protein complexes followed by mass- spectrometric identification of the components. Most proteins are thought to exist in the cell not as free entities but as part of ï¿½cellular machines' which perform cellular functions cooperatively. Systematic identification of protein complexes would permit these machines to be defined and allow ï¿½physical maps' to be created for a variety of cell types and states. Such information is of great value for the assignment of protein function.

Cell maps: A cell map specifies the proteins that constitute a given organelle within a given cell type. Cell maps for normal and diseased cells can be constructed which give insight into the role proteins have in disease and can guide the drug development process.

Centromere: A region of a chromosome to which spindle traction fibers attach during mitosis and meiosis; the position of the centromere determines whether the chromosome is considered an acrocentric, metacentric or telomeric chromosome.

Charcot-Marie Tooth disease: A condition characterized by degeneration of the motor and sensory nerves that control movement and feeling in the arm below the elbow and the leg below the knee; transmitted in autosomal dominant, autosomal recessive and X-linked forms.

Chemical proteomics: To link new proteins with known catalytic activities, proteome-scale screens for generic enzyme activities (e.g. protease and phosphatase) should be implemented ... Although it is impossible to screen for chemical reactions that are unknown, in theory, identifying small molecules that bind to the new proteins may elucidate clues to new activities.

Chorionic villus sampling: An invasive prenatal diagnostic procedure involving removal of villi from the human chorion to obtain chromosomes and cell products for diagnosis of disorders in the human embryo.

Chromosomal maps: Genes or other identifiable DNA fragments are assigned to their respective chromosomes, with distances measured in base pairs. These markers can be physically associated with particular bands (identified by cytogenetic staining) primarily by in situ hybridization, a technique that involves tagging the DNA marker with an observable label (e.g., one that fluoresces or is radioactive). The location of the labeled probe can be detected after it binds to its complementary DNA [cDNA] strand in an intact chromosome.

Chromosome banding: A technique for staining chromosomes so that bands appear in a unique pattern particular to the chromosome.

Chromosome mapping: Any method used for determining the location of and relative distances between genes on a chromosome.

Chromosome: In the eukaryotic nucleus, one of the threadlike structures consisting of chromatin and carry genetic information arranged in a linear sequence.

Cleft lip/palate: Congenital condition with cleft lip alone, or with cleft palate; cause is thought to be multifactorial.

Clone: Genetically engineered replicas of DNA sequences.

Clone-based maps: The physical map of the human genome published by Nature is a clone- based physical map of 3.2 gigabases (25 times larger than any previously mapped genome). This approach involved generating an overlapping series of clones for the whole genome. With a fingerprinted BAC map clones could be selected for sequencing ensuring comprehensive coverage of the genome.

Cloned DNA: Any DNA fragment that passively replicates in the host organism after it has been joined to a cloning vector.

Codon: A sequence of three nucleotides in mRNA that specifies an amino acid.

Comparative genome mapping: Comparative genome mapping in the sequence-based era: early experience with human chromosome 7

Comparative proteomics: The *C. elegans* proteome was used as an alignment template to assist in novel human gene identification ... Among the available 18,452 *C. elegans* protein sequences, our results indicate that at least 83% had human homologous genes, with 7954 records of *C. elegans* proteins matching known human gene transcripts.

Computational proteomics: Large- scale generation and analysis of 3D and 4D protein structural information and the application of structural knowledge across all life science disciplines.

Consanguinity: Genetic relationship. Consanguineous individuals have at least one common ancestor in the preceding few generations.

Conservative change: An amino acid change that does not affect significantly the function of the protein.

Contig mapping: Overlapping of cloned or sequenced DNA to construct a continuous region of a gene, chromosome or genome.

Contig maps: Contig maps are important because they provide the ability to study a complete, and often large segment of the genome by examining a series of overlapping clones which then provide an unbroken succession of information about that region.

Contiguous genes: Genes physically close on a chromosome that when acting together express a phenotype.

Cornelia de Lange syndrome: Condition involving growth deficiency, significant developmental delay, anomalies of the extremities and a characteristic facial appearance.

Cosmid maps: "Constructing chromosome- and region-specific cosmid maps of the human genome"

Cosmids: Plasmid vectors designed for cloning large fragments of eukaryotic DNA;

the vector is a plasmid into which phage lambda cohesive end sites have been inserted.

CpG islands: Areas of multiple CG repeats in DNA.

Cri-du-chat syndrome: A chromosomal condition (monosomy 5p). Name comes from the distinctive mewing cry of affected infants; characterized by significant mental deficiency, low birthweight, failure to thrive and short stature; deletion of a small section of the short arm of chromosome 5.

Crossovers: The exchange of genetic material between two paired chromosome during meiosis.

Cystic fibrosis: An autosomal recessive genetic condition of the exocrine glands, which causes the body to produce excessively thick, sticky mucus that clogs the lungs and pancreas, interfering with breathing and digestion.

Cytogenetic maps: The visual appearance of a chromosome when stained and examined under a microscope. Particularly important are visually distinct regions, called light and dark bands, which give each of the chromosomes a unique appearance. This feature allows a person's chromosomes to be studied in a clinical test known as a karyotype, which allows scientists to look for chromosomal alterations.

Degenerate codon: A codon that specifies the same amino acid as another codon.

Deletion mapping: The use of overlapping deletions to localize the position of an unknown gene on a chromosome or linkage map.

Deletion: The loss of a segment of the genetic material from a chromosome.

Disease: Any deviation from the normal structure or function of any part, organ, or system of the body that is manifested by a characteristic set of symptoms and signs whose pathology and prognosis may be known or unknown.

DNA fingerprint technique: A method employed to determine differences in amino acid sequences between related proteins; relies upon the presence of a simple tandem-repetitive sequences that are scattered throughout the human genome.

DNA hybridization: A technique for selectively binding specific segments of single-stranded (ss) DNA or RNA by base pairing to complementary sequences on ssDNA molecules that are trapped on a nitrocellulose filter.

DNA probe: Any biochemical used to identify or isolate a gene, a gene product, or a protein.

DNA sequencing: "Plus and minus" or "primed synthesis" method, developed by Sanger, DNA is synthesized in vitro in such a way that it is radioactively labeled and the reaction terminates specifically at the position corresponding to a given base; the "chemical" method, ssDNA is subjected to several chemical cleavage protocols that selectively make breaks on one side of a particular base.

Dominant: Alleles that determine the phenotype displayed in a heterozygote with another (recessive) allele.

Down syndrome: A type of mental deficiency due to trisomy (three copies) of autosome 21, a translocation of 21 or mosaicism.

Duchenne/Becker muscular dystrophy: The most common and severe form of muscular dystrophy; transmitted as an X-linked trait. X-linked recessive. Symptoms include onset at 2-5 years with difficulty with gait and stairs, enlarged calf muscles, progression to wheelchair by adolescence, shortened life span.

Dwarfism: Conditions of short stature with adult height under 4'10" as adult, usually with normal intelligence and lifespan. Ehlers Danlos Syndrome connective tissue condition including problems with tendons, ligaments, skin, bones, cartilage, and membranes surrounding blood vessels and nerves. Symptoms include joint laxity, elastic skin, dislocations. Many forms: autosomal dominant, autosomal recessive, X-linked forms.

Dystonia: Neurologic condition involving repeated twisting and movement. Involves a variety of muscle groups. Intelligence not effected. Three forms: childhood - autosomal dominant, autosomal recessive, adult-acquired.

ELSI: Ethical, legal and social implications (of HGP).

Endonuclease: An enzyme that breaks the internal phosphodiester bonds in a DNA molecule.

Epitope mapping: Methods used for studying the interactions of antibodies with specific regions of protein antigens. Important applications of epitope mapping are found within the area of immunochemistry.

Erythrocytes: The hemoglobin-containing cell found in the blood of vertebrates.

Ethics: The study of fundamental principles which defines values and determines moral duty and obligation.

Euchromatin: The chromatin that shows the staining behavior characteristic of the majority of the chromosomal complement.

Eugenics: The improvement of humanity by altering its genetic composition by encouraging breeding of those presumed to have desirable genes.

Evolutionary genetics: Evolutionary study of genes has been purely theoretical, but it can provide useful information for guiding gene mapping. People are now finding, for example, that a lot of things are not true associations;

instead, they are an artifact of association. You can make such mistakes when you are looking at two individuals who share a common ancestry. Understanding the phylogeny helps us, for example, understand horizontal gene transfer between microorganisms. For humans or other sexually reproducing organisms, the use of phylogenetic information improves resolution for making associations by helping to avoid type I errors - that is, finding an association that is actually merely due to sharing a recent common ancestor, or, in other words, being closely related.

Exons: Portion of a gene included in the transcript of a gene and survives processing of the RNA in the cell nucleus to become part of a spliced messenger of a structural RNA in the cell cytoplasm; an exon specifies the amino acid sequence of a portion of the complete polypeptide.

Expressed Protein Tags EPTs: Multi- cellular organisms have been evolving a system with which they can discriminate between cells of their own origin and other adventitious cells or cells which have been infected with intracellular pathogens. To achieve this goal, a family of receptors, known as multi- ligand receptors (MLR), have evolved to be remarkably promiscuous binders of peptide ligands. The MLR-bound ligands are derived from degradation intermediates of cellular proteins. Typically, these ligands are 8-12 amino acids in length and have been coined by CANVAS as "expressed protein tags" or EPTs. EPTs are of sufficient length to differentiate particular proteins and/ or individual genes. Each MLR has a single binding site and thus contains a single EPT copy.

Expression imbalance map EIM: A new visualization method, for detecting mRNA expression imbalance regions, reflecting genomic losses and gains at a much higher resolution than conventional technologies such as comparative genomic hybridization (CGH). Simple spatial mapping of the microarray expression profiles on chromosomal location provides little information about genomic structure, because mRNA expression levels do not completely reflect genomic copy number and some microarray probes would be of low quality. The EIM, which does not employ arbitrary selection of thresholds in conjunction with hypergeometric distribution- based algorithm, has a high tolerance of these complex factors.

Expression mapping: The creation of quantitative maps of protein expression from cell or tissue extracts, akin to the EST maps commercially available. This approach relies on 2D gel maps and image analysis, and opens up the possibility of studying cellular pathways and their perturbation by disease, drug action or other biological stimuli at the whole- proteome level ï¿½ Expression mapping is a valuable tool in the discovery of disease markers and its use in gaining information in toxicological and drug- action studies seems assured. It is unclear at present how successful this approach will be in elucidating cellular pathways and their importance in disease processes, and how much the precise measurement of protein levels matters when compared with the rough guide provided by the measurement of mRNA levels ï¿½ the ability to measure protein- level changes directly would seem to carry inherent advantages and it seems likely that expression proteomics will be a useful tool in drug target discovery and in studying the effects of various biological stimuli on the cell.

Fetal alcohol syndrome: A link between excessive alcohol consumption during pregnancy and birth defects; characteristics include small head and eyes, folds of the skin that obscure the inner juncture of the eyelids, short, upturned nose, and thin lips.

FISH: Florescent in situ hybridization: a technique for uniquely identifying whole chromosomes or parts of chromosomes using florescent tagged DNA.

Fragile sites: A non-staining gap of variable width that usually involves both chromatids and is always at exactly the same point on a specific chromosome derived from an individual or kindred.

Fragile-X syndrome: X-linked trait; the second most common identifiable cause of genetic mental deficiency.

Functional glycomics: At the present time approximately 110 glycogenes which encode glycosyltransferases and related genes have been cloned. Even though some of the biological functions of those genes have been elucidated, most of the actual functions of these genes continue to be obscure, and limited information is available in terms of their pathophysiolgical significance. Therefore in order to clarify the functional significance of these genes, one of the major strategies is focused on the identification of likely target molecules in vivo and the identification of their functional significance. **Knock out** or **transgenic** mice have already been reported for several glycogenes and currently available information indicate that some are lethal and some lead to interesting phenotypic changes.

Functional maps: In addition to the raw data, it will be important to design the proper visualization tools to graphically represent the functional relationships contained in different maps ... Finally, it will be important to consider the possibility that functional maps need to be related back to particular tissues or even cell types.

Gamete: An haploid cell.gel electrophoresis the process by which nucleic acids (DNA or RNA) or proteins are separated by size according to movement of the charged molecules in an electrical field.

Gene amplification: Any process by which specific DNA sequences are replicated disproportionately greater than their representation in the parent molecules; during development, some genes become amplified in specific tissues.

Gene map: The linear arrangement of mutable sites on a chromosome as deduced from genetic recombination experiments.

Gene mapping: Determination of the relative positions of genes on a DNA molecule (chromosome or plasmid) and of the distance, in linkage units or physical units, between them. [DOE]

Gene therapy: Addition of a functional gene or group of genes to a cell by gene insertion to correct an hereditary disease.

Gene: A hereditary unit that occupies a certain position on a chromosome; a unit that has one or more specific effects on the phenotype, and can mutate to various allelic forms.

Genetic counseling: The educational process that helps individuals, couples, or families to understand genetic information and issues that may have an impact on them.

Genetic linkage map: A chromosome map showing the relative positions of the known genes on the chromosomes of a given species.

Genetic maps: Also known as a linkage map. A chromosome map of a species that shows the position of its known genes and/ or markers relative to each other, rather than as specific physical points on each chromosome. The value of the genetic map is that an inherited disease can be located on the map by following the inheritance of a DNA marker present in affected individuals (but absent in unaffected individuals), even though the molecular basis of the disease may not yet be understood nor the responsible gene identified. Genetic maps have been used to find the exact chromosomal location of several important disease genes, including cystic fibrosis, sickle cell disease, Tay- Sachs disease, fragile X syndrome, and myotonic dystrophy.

Genetic screening: Testing groups of individuals to identify defective genes capable of causing hereditary conditions.

Genetic variation: A phenotypic variance of a trait in a population attributed to genetic heterogeneity.

Genome control maps: Would identify all the components of the transcriptional machinery that have roles at any particular promoter and the contribution that specific components make to coordinate regulation of genes. The map will facilitate modeling of the molecular mechanisms that regulate gene expression and implicate components of the transcription apparatus in functional interactions with gene-specific regulators.

Genome fingerprint map: The collection of all fingerprint clone contigs placed in a genome- wide map.

Genome map: A reconstruction of the entire set of chromosomes for a given organism, showing the relative position of every gene.

Genome scale metabolic maps: Annotated genomic data, along with legacy data on the cell's biochemistry and physiology, can be used to construct genome-scale metabolic maps. The challenge now is to formulate reliable mathematical descriptions of the integrated function of these maps. It has proven difficult, if not impossible, to formulate detailed theory-based models of these genome-scale maps. An alternative approach that is data driven and constraints based will be described. It is an iterative model-building process.

Genome: All of the genes carried by a single gamete; the DNA content of an individual, which includes all 44 autosomes, 2 sex chromosomes, and the mitochondrial DNA.

Genomic cartography: [Fry's] research focuses on methods of visualizing large amounts of data from dynamic information sources. The work uses ideas from distributed and adaptive systems to form organic representations that react and respond to the input data. This work is currently directed towards Genomic Cartography which is a study into new methods to represent the data found in the human genome.

Genomic mapping: While a few technologies for functional analysis on a genomic basis are being developed at present, additional approaches and technologies for genomic interpretation that can be applied efficiently and economically at the level of an entire genome will be required for comprehensive analyses. Informatics will continue to play an important role in achieving all of these goals, as well as in ensuring the maintenance and accessibility of the forthcoming data. The development and application of new technologies for acquisition, management, analysis, and dissemination of genomic data are still required.

Genotype: Genetic constitution of an organism.

Germ cell: A sex cell or gamete (egg or spermatozoan).Haldane equation Haldane's law: the generalization that if first generation hybrids are produced between two species, but one sex is absent, rare, or sterile, that sex is the heterogamic sex.

Haplotype map: Francis Collins, director of the NHGRI, speaking at BIO 2001 (San Diego CA, US, June 2001) announced plans for a public- private effort to create a human haplotype map. Creators hope this so- called haplotype map will be a tool for pinning down the genes that contribute to the development of complex diseases such as cancer, diabetes, and mental illness.

Haplotype mapping: Is often carried out as part of a genome scan. In a population isolate, the appearance of a rare Mendelian disease is almost always attributable to a single founder gene or mutation. The disease allele can be identified by searching for a common haplotype signature shared among patients. As the ancestral haplotype signature is passed from generation to generation, it is disrupted by recombination. Partial conservation of the haplotype signature in a patient strongly suggests that the disease locus resides in the conserved region of the haplotype.

Hardy-Weinberg Law: The concept that both gene frequencies and genotype frequencies will remain constant from generation to generation in an infinitely large, interbreeding population in which mating is at random and there is no selection, migration or mutation.

Hemophilia: A sex-linked disease in humans in which the blood-clotting process is defective.

Heterogeneity: The production of identical or similar phenotypes by different genetic mechanisms.

High- resolution physical mapping: The two current approaches are termed top-down (producing a macrorestriction map) and bottom- up (resulting in a contig map). With either strategy the maps represent ordered sets of DNA fragments that are generated by cutting genomic DNA with restriction enzymes. The fragments are then amplified by cloning or by polymerase chain reaction (PCR) methods. Electrophoretic techniques are used to separate the fragments according to size into different bands, which can be visualized by direct DNA staining or by hybridization with DNA probes of interest. The use of purified chromosomes separated either by flow sorting from human cell lines or in hybrid cell lines allows a single chromosome to be mapped.

High-resolution genetic maps: 2-5 cM [centiMorgans]. Genetic mapping resolution has been increased through the application of recombinant DNA technology, including in vitro radiation- induced chromosome fragmentation and cell fusions (joining human cells with those of other species to form hybrid cells) to create panels of cells with specific and varied human chromosomal components.

Homologous chromosomes: Chromosomes that pair during meiosis; each homologue is a duplicate of one chromosome from each parent.

Homology map: The Davis Human/ Mouse Homology Map, a table comparing genes in homologous segments of DNA from human and mouse sources, sorted by position in each genome. A total of 1793 loci are presented, most of which are genes. The authors did not include pseudogenes, members of multigene families where specific homology relationships could not be determined, nor any other genes for which homology was in doubt. In addition, for 568 of the loci there are provisional assignments of markers that link the homology map with that of the Gene Map of the Human Genome. . These links also provide a rough approximation of the position of markers in the Genethon linkage map. In constructing this table, the authors first ordered genes so as to best maintain order according to both human cytogenetic position and mouse genetic map position. Within these homologous regions, genes were ordered according to the mouse genetic mapping data.

Homozygote: Having identical alleles at one or more loci in homologous chromosome segments.

Housekeeping genes: Those genes expressed in all cells because they provide functions needed for sustenance of all cell types.

HUGO: Human Genome Organization.

Human Proteome Organisation HUPO: The reason for creating HUPO is to assist in increasing the awareness of this discipline of science across society, particularly with regard to the Human Proteome Project and to engender a broader understanding of the importance of proteomics and the opportunities it offers in the diagnosis, prognosis and therapy of disease. As a global body it will also have the objective of fostering international cooperation across the proteomics community and of promoting scientific research in an on- going manner around the world..

Huntington disease: A disease characterized by irregular, spasmodic involuntary movements of the limbs and facial muscles, mental deterioration and death, usually within 20 years of the onset of symptoms.

Hybridization: The pairing of a single-stranded, labeled probe (usually DNA) to its complementary sequence.

Ichthyosis: Any of several hereditary or congenital skin conditions; skin of affected individuals has a dry, scaly appearance.

Imprinting: A chemical modification of a gene allele which can be used to identify maternal or paternal origin of chromosome.

Incomplete penetrance: The gene for a condition is present, but not obviously expressed in all individuals in a family with the gene.

Interaction proteomics: Protein- protein interactions lie at the heart of most cellular processes ... A complete understanding of cellular function depends on a full characterization of the complex network of cellular protein- protein associations .... Alternative proteomics technologies are being developed to complement the two- hybrid system. These methods reveal direct protein-protein interactions by using protein affinity chromatography. Protein affinity chromatography, as developed by Greenblatt, Alberts, and colleagues, has the disadvantage of requiring purified proteins as reagents, but it is superior to the two- hybrid approach because it generates fewer false positives and is more amenable to high- throughput screening.

International SNP Map Working Group: Identifies and localizes 1.42 millions SNPs in the human genome. ["A map of human genome sequence variation containing 1.42 million single nucleotide polymorphisms".

Interologs: Protein interaction maps have provided insight into the relationships among the predicted proteins of model organisms for which a genome sequence is available. These maps have been useful in generating potential interaction **networks**, which have confirmed the existence of known complexes and **pathways** and have suggested the existence of new complexes and or crosstalk between previously unlinked pathways. However, the generation of such maps is costly and labor intensive. Here, we investigate the extent to which a protein interaction map generated in one species can be used to predict interactions in another species.

Introns: A segment of DNA (between exons) that is transcribed into nuclear RNA, but are removed in the subsequent processing into mRNA.

Isochromosome: A metacentric chromosome produced during mitosis or meiosis when the centromere splits transversely instead of longitudinally; the arms of such chromosome are equal in length and genetically identical, however, the loci are positioned in reverse sequence in the two arms.

Karyotype: A set of photographed, banded chromosomes arranged in order from largest to smallest.

Klinefelter syndrome: An endocrine condition caused by a an extra X-chromosome (47,XXY); characterized by the lack of normal sexual development and testosterone, leading to infertility and adjustment problems if not detected and treated early.

Linkage disequilibrium: Evidence for linkage disequilibrium can be helpful in mapping disease genes since it suggests that the two [alleles] may be very close to one another

Linkage maps: A map of the relative positions of genetic loci on a chromosome, determined on the basis of how often the loci are inherited together. Distance is measured in centimorgans (cM).

Linkage: Analysis of pedigree the tracking of a gene through a family by following the inheritance of a (closely associated) gene or trait and a DNA marker.

Linkage: The greater association in inheritance of two or more nonallelic genes than is to be expected from independent assortment; genes are linked because they reside on the same chromosome.

Lligase: An enzyme that functions in DNA repair.

Localizome mapping: One can imagine comprehensive mapping projects of the "localizome", with the goal of recording not only where all proteins of a proteome can be found but also when.

Locus: Any genomic site, whether functional or not, that can be mapped through formal genetic analysis.

Lod score: logarithm of the odd score; a measure of the likelihood of two loci being within a measurable distance of each other.

Macrorestriction map: Describes the order and distance between enzyme cutting (cleavage) sites ... In top- down mapping, a single chromosome is cut (with rare- cutter restriction enzymes) into large pieces, which are ordered and subdivided; the smaller pieces are then mapped further. The resulting macro-restriction maps depict the order of and distance between sites at which rare-cutter enzymes cleave. This approach yields maps with more continuity and fewer gaps between fragments than contig maps, but map resolution is lower and may not be useful in finding particular genes; in addition, this strategy generally does not produce long stretches of mapped sites. Currently, this approach allows DNA pieces to be located in regions measuring about 100,000 bp to 1 Mb.

Mapping: The determination of the relative positions of genes within the chromosomes or of restriction sites along a DNA molecule.The process of determining the position of a locus on the chromosome relative to other loci.

Marfan syndrome: Autosomal dominant condition of connective tissue; affects the skeletal, ocular and cardiovascular systems.

Marker: A gene with a known location on a chromosome and a clear-cut phenotype, used as a point of reference when mapping a new mutant.

Meiosis: The doubling of gametic chromosome number.

Methylation: Addition of a methyl group (-CH3) to DNA or RNA.

Methylmalonic acidemia: A group of conditions characterized by the inability to metabolize methylmalonic acid or by a defect in the metabolism of Vitamin B12.

Microbial proteomics: Bacterial genomes encode all possible virulence determinants, vaccine candidates, and potential drug targets. Further, a completed genomic sequence establishes a basis for high throughput analysis of the proteins expressed (i.e., the proteome). Respiratory pathogens have been among the first to have their genomes entirely sequenced.

Missense mutation: A change in the base sequence of a gene that alters or eliminates a protein.

Mitochondrial DNA: The mitochondrial genome consists of a circular DNA duplex, with 5 to 10 copies per organelle.

Multifactorial: A characteristic influenced in its expression by many factors, both genetic and environmental.

Mutation: Process by which genes undergo a structural change.

Myotonic dystrophy: A combination of progressive weakening of the muscles and muscle spasms or rigidity, with difficulty relaxing a contracted muscle; inherited as an autosomal dominant trait.

Neurofibromatosis: One of the most common single gene conditions affecting the human nervous system; in most cases, "cafe au lait" spots, are the only symptom; inherited as an autosomal dominant trait, with 50% being new mutations.

Nonsense mutation: A mutation in which a codon is changed to a stop codon, resulting in a truncated protein product.

Noonan syndrome: A condition characterized by short stature and ovarian or testicular dysfunction, mental deficiency, and lesions of the heart.

Northern analysis: A technique for transferring electrophoretically resolved RNA segments from an agarose gel to a nitrocellulose filter paper sheet via capillary action.

Nucleotide mapping: Two- dimensional separation and analysis of nucleotides.

Nucleotide: One of the monomeric units from which DNA or RNA polymers are constructed; consists of a purine or pyrimidine base, a pentose sugar and a phosphoric acid group.

Oncogenes: Genes involved in cell cycle control (growth factors, growth factor regulator genes, etc), a mutation can lead to tumor growth.

Optical mapping: An enabling technology for whole genome analysis which involves the capture of individual DNA molecules, obtained directly from genomic DNA, followed by digestion in situ by selected restriction endonucleases. The resulting fragments are then visualized directly to produce detailed optical restriction maps. This methodology allows patterns of sequence variation to be detected across entire genomes, without the need for DNA amplification, and, unlike other genomewide scanning methods, provides detailed haplotype information by analyzing individual DNA molecules. OpGen will provide optical mapping services to three main markets ï¿½ i.e., genome sequencing projects, cancer diagnostics, and genetic association studies.

Osteogenesis imperfecta: A condition also known as brittle bone disease; characterized by a triangular shaped face with yellowish brown teeth, short stature and stunted growth, scoliosis, high pitched voice, excessive sweating and loose joints.

Parthenogenesis: The development of an individual from an egg without fertilization.

PCR: Polymerase chain reaction; a technique for copying the complementary strands of a target DNA molecule simultaneously for a series of cycles until the desired amount is obtained.

Pedigree: A diagram of the heredity of a particular trait through many generations of a family.

Peptide mapping: Two- dimensional separation and analysis of peptides.The characteristic pattern of fragments formed by the separation of a mixture of peptides resulting from hydrolysis of a protein or peptide.

Peptide maps: One way to research the structure of a protein is to seek to identify the specific sequence of amino acids which form the protein. Tr. at 202-3. An initial approach is to cut the protein into fragments, called peptides, using enzymes or chemicals which reliably divide the chain at predictable points. Tr. at 523. These fragments can be isolated and analyzed to create "peptide maps," which show the pattern of pieces from these breaks as a unique "fingerprint." ... A peptide map does not disclose the precise order of the amino acids in a protein, although it may point to areas of difference between similar proteins.

Phenome mapping: The conceptual matrix for a comprehensive "phenome" mapping project would be as follows: one axis represents all available knockouts while the other represents a large series of standardized phenotypes that can be screened.

Phenome maps: Can be thought of as lists of similar phenotypes that could be referred to as "pheno- clusters".

Phenotype: Observable characteristics of an organism produced by the organism's genotype interacting with the environment.

Physical Map: Map where the distance between markers is the actual distance, such as the number of base pairs.

Physical Mapping: The procedure of physical mapping coarsely divides into two steps First large pieces of DNA (contigs), a library of cloned fragments are ordered according to their position in the genome. Different experimental techniques are used to do that. Roughly, these are clone- probe hybridization mapping, restriction mapping, radiation- hybrid mapping and optical mapping. ... Second the cloned fragments are cut by restriction enzymes, smaller DNA fragments are obtained which are sequenced in detail (shotgun- sequencing), and the overall sequence in detail is obtained by Sequence Assembly.

Physical Maps: A map of the locations of identifiable landmarks on DNA (e.g. restriction enzyme cutting sites, genes) regardless of inheritance. Distance is measured in base pairs. For the human genome, the lowest- resolution physical map is the banding patterns on the 24 different chromosomes; the highest resolution map would be the complete nucleotide sequence of the chromosomes. A chromosome map of a species that shows the specific physical locations of its genes and/ or markers on each chromosome. Physical maps are particularly important when searching for disease genes by positional cloning strategies and for DNA sequencing.

PKU: phenylketonuria, an enzyme deficiency condition characterized by the inability to convert one amino acid, phenylalanine, to another, tyrosine, resulting in mental deficiency. plasmid double-stranded, circular, bacterial DNA into which a fragment of DNA from another organism can be inserted.

Pleiotropy: The phenomenon of variable phenotypes for a number of distinct and seemingly unrelated phenotypic effects.

Polycystic Kidney Disease (PKD): A group of conditions characterized by fluid filled sacs that slowly develop in both kidneys, eventually resulting in kidney malfunction.

Polymerase: Any enzyme that catalyzes the formation of DNA or RNA from deoxyribonucleotides or ribonucleotides.

Positional Cloning: Requires a genetic map with a large number of markers (especially in the region of interest), and the use of physical mapping and DNA sequencing technologies to isolate and sequence the targeted gene.

Prader-Willi Syndrome: A condition characterized by obesity and insatiable appetite, mental deficiency, small genitals, and short stature. May be deletion of #15 chromosome.

Predisposition: To have a tendency or inclination towards something in advance.

Presymptomatic Diagnosis: diagnosis of a genetic condition before the appearance of symptoms.

Primer: Nucleotides used in the polymerase chain reaction to initiate DNA synthesis at a particular location.

Probability: The long term frequency of an event relative to all alternative events, and usually expressed as decimal fraction.

Proband: Individual in a family who brought the family to medical attention.

Probe: Single-stranded DNA labeled with radioactive isotopes or tagged in other ways for ease in identification.

Prognosis: Prediction of the course and probable outcome of a disease.

Protein and mRNA Data: Although the relationship between mRNA and protein levels is vague for individual genes, some of the statistics for broad categories of protein properties are much more robust... In contrast to the differences between mRNA and protein data for individual genes, the broad categories show that the **transcriptome** and **translatome** populations are remarkably similar; both contain roughly the same proportions of secondary structure and functional categories. Moreover, this contrasts the difference with the *genome,* which appears to have a distinctly different composition of functional categories. This illustrates that we get a more consistent picture when we average across the population, i.e. there is broad similarity between the characteristics of highly expressed mRNA and highly abundant proteins.

Protein- carbohydrate interactions: Are now recognized to be important mediators of cell communication. In the last decade many novel carbohydrate binding proteins (CBPs) have been described, and several have been documented to play critical roles in cell trafficking and cell signaling. Despite these advances, the rate of generating new information has been slow, and the biological roles of most mammalian CBPs remain poorly understood. While the importance of this field has attracted many outstanding laboratories, a major barrier to rapid progress has been the structural complexity and heterogeneity of the carbohydrates themselves, and the analytical and synthetic challenges this poses. A further complexity arises from the fact that carbohydrate ligands are post-translational modifications of proteins and lipids whose synthesis is directed by the coordinated expression of multiple genes in a manner that is not template driven.

Protein databases: Dr. Stanley Fields, Professor of Genetics and Medicine at the Univ. of Washington and developer of the yeast two hybrid system writes that protein databases "will need to become much more sophisticated if they are to help scientists make sense of the staggering number of experimental measurements that will soon emerge. ... protein data will need to be integrated with results from expression profiling, genome- wide mutation or antisense analyses, and polymorphism detection. As proteomic data accumulate, we will become better at triangulating from multiple disparate bits of information to gain a bearing on what a protein does in the cell.

Protein Expression Map: Since 2D Electrophoresis gel patterns reveal not only the amounts of protein, but is unrivaled in its ability to detect post- translational modifications, the 2DE protein map provides much more relevant information about cellular dynamics than the corresponding expression map at the mRNA level. By comparing the 2DE gel patterns of samples exposed to different physiological conditions or different drug treatments it is possible to identify groups of proteins with related functions or whose expression is interdependent (expression proteomics).

Protein Expression Mapping PEM: Details the distribution and abundance of protein in specific samples, under defined physiological conditions. [CHI Proteomics] Quantitative study of global changes in protein expression in tissues, cells or body fluids using 2D gels and image analysis. Currently carried out by 2D gel electrophoresis, though alternatives are under investigation.

Protein Function: In the expanded view of protein function, a protein is defined as an element in the network of its interactions. Various terms have been coined for this expanded notion of function, such as 'contextual function' or 'cellular function' ... Whatever the term, the idea is that each protein in living matter functions as part of an extended web of interacting molecules ... Often it is possible to understand the cellular functions of uncharacterized proteins through their linkages to characterized proteins. In broader terms, the networks of linkages offer a new view of the meaning of protein function, and in time should offer a deepened understanding of the function of cells.

Protein interaction maps: Hybrigenics" comprehensive protein interaction maps using automated yeast- two- hybrid methodology in pathogens and in cDNA of normal and diseased tissues.

Protein linkage maps: With respect to a genome- wide use of the two- hybrid assay in the case of yeast, the goal is to find which proteins in the yeast genome interact with every other protein. This process would generate protein linkage maps, delineating large networks of interacting proteins. The approximately 6,000 yeast proteins can potentially interact in 18 million pairwise combinations.

Proteome map: A number of organizations have announced plans to produce a map of the proteome, including Myriad Genetics, Large Scale Biology, CuraGen and others.

Proteomics: The analysis of complete complements of proteins. Proteomics includes not only the identification and quantification of proteins, but also the determination of their localization, modifications, interactions, activities, and, ultimately, their function. Initially encompassing just two- dimensional **(2D) gel** *electrophoresis* for protein separation and identification, proteomics now refers to any procedure that characterizes large sets of proteins. The explosive growth of this field is driven by multiple forces - *genomics* and its revelation of more and more new proteins; powerful protein technologies, such as newly developed *mass spectrometry* approaches, global [yeast] **two- hybrid** techniques, and spin- offs from **DNA arrays**; and innovative computational tools and methods to process, analyze, and interpret prodigious amounts of data.

Proteus syndrome: A condition characterized by distorted asymmetric growth of the body and enlarged head, enlarged feet, multiple nevi on the skin; mode of inheritance is unknown.

Public policy: A set of action guidelines or rules that result from the actions or lack of actions of governmental entities.

QTL mapping Quantitative Trait Loci mapping: A phenotype driven approach to gene function. As such it permits the discovery of new genes and can be contrasted with gene- driven approaches such knock-out and knock-in mice which allow for the study of known genes. QTL reflect natural genetic variations as they exist in the mouse strains under study. We are limited to detecting those genes that vary among the available strains. However the natural variations among mouse strains are vast and largely untapped.

Radiation hybrid mapping: A method for ordering genetic loci along CHROMOSOMES. The method involves fusing irradiated donor cells with host cells from another species. Following cell fusion, fragments of DNA from the irradiated cells become integrated into the chromosomes of the host cells. Molecular probing of DNA obtained from the fused cells is used to determine if two or more genetic loci are located within the same fragment of donor cell DNA.

Radiation Hybrid RH maps: Chromosome maps that are calculated from RH score vectors. An RH score vector is the pattern of assay results of a particular STS (marker) on a particular panel. The vector consists of 1's (did amplify) and 0's (did not amplify). Simplistically speaking, the more similar two score vectors are, the closer the markers are on the chromosome.

Recessive: A gene that is phenotypically manifest in the homozygous state but is masked in the presence of a dominant allele.

Recombination: The natural process of breaking and rejoining DNA strands to produce new combinations of genes and, thus, generate genetic variation. Gene crossover during meiosis.

Repeat sequences: The length of a nucleotide sequence that is repeated in a tandem cluster.

Restriction map: A description of restriction endonuclease cleavage sites within a piece of DNA. Generating such a map is usually the first step in characterizing an unknown DNA, and a prerequisite to manipulating it for other purposes. Typically, restriction enzymes that cleave DNA infrequently (e.g. those with 6 bp recognition sites) and are relatively inexpensive are used to produce at a map.

Restriction mapping: Use of restriction endonucleases to analyze and generate a physical map of genomes, genes, or other segments of DNA.

Retinitis pigmentosa: Group of hereditary ocular disorders with progressive retinal degeneration. Autosomal dominant, autosomal recessive, and x-linked forms.

Retinoblastoma: A childhood malignant cancer of the retina of the eye. reverse transcriptase viral enzyme used to make cDNA.

Reverse-two hybrid: A variation of the **yeast two hybrid** system, in which protein- protein interactions increase the transcription of a toxic counterselectable marker, resulting in growth inhibition. The availability of a counterselectable marker significantly extends the possibilities of the two- hybrid system. Most importantly, dissociation of **protein- protein interactions** can be selected for, and thus protein- protein interactions can be characterized and manipulated genetically.

RFLP (Restriction Fragment Length Polymorphism): See Genetic variations glossary Polymorphic sequences that result in RFLPs are used as markers on both physical maps and genetic linkage maps. RFLPs are usually caused by mutation at a cutting site.

RFLP: Restriction fragment length polymorphism; variations occurring within a species in the length of DNA fragments generated by a species endonuclease.

Ribosomal protein: One of the ribonucleoprotein particles that are the sites of translation.

Rubinstein-Taybi syndrome: Condition with multiple congenital anomalies including: mental deficiency, broad thumbs, small head, broad nasal bridge and beaked nose.

Sanger sequence: "Plus and minus" or "primed synthesis" method; DNA is synthesized so it is radioactively labeled and the reaction terminates specifically at the position corresponding to a given base.

Selection: The process of determining the relative share allotted individuals of different genotypes in the propagation of a population; the selective effect of a gene can be defined by the probability that carriers of the gene will reproduce.

Sex determination: The mechanism in a given species by which sex is determined; in many species sex is determined at fertilization by the nature of the sperm that fertilizes the egg.

Sickle cell anemia: An hereditary, chronic form of hemolytic anemia characterized by breakdown of the red blood cells; red blood cells undergo a reversible alteration in shape when the oxygen tension of the plasma falls slightly and a sickle-like shape forms.

SNP maps: A collection of SNPs that can be superimposed over the existing genome map, creating greater detail and facilitating further genetic studies. Current estimates indicate that a very dense marker map (30,000 - 1,000,000 variants) would be required to perform haplotype - based association studies. We have constructed a SNP map of the human genome with sufficient density to study human haplotype structure, enabling future study of human medical and population genetics.

Somatic cell hybrid: Hybrid cell line derived from two different species; contains a complete chromosomal complement of one species and a partial chromosomal complement of the other; human/hamster hybrids grow and divide, losing human chromosomes with each generation until they finally stabilize, the hybrid cell line established is then utilized to detect the presence of genes on the remaining human chromosome.

Somatic mutation: A mutation occurring in any cell that is not destined to become a germ cell; if the mutant cell continues to divide, the individual will come to contain a patch of tissue of genotype different from the cells of the rest of the body.

Southern blotting: A technique for transferring electrophoretically resolved DNA segments from an agarose gel to a nitrocellulose filter paper sheet via capillary action; the DNA segment of interest is probed with a radioactive, complementary nucleic acid, and its position is determined by autoradiography.

Spina bifida: A congenital condition that results from altered fetal development of the spinal cord, part of the neural plate fails to join together and bone and muscle are unable to grow over this open section.

Syndrome: A recognizable pattern or group of multiple signs, symptoms or malformations that characterize a particular condition; syndromes are thought to arise from a common origin and result from more than one developmental error during fetal growth.

Tay-Sachs disease: A fatal degenerative disease of the nervous system due to a deficiency of hexosamidase A, causing mental deficiency, paralysis, mental deterioration, and blindness; found primarily but not exclusively among Ashkenazi Jews. Autosomal recessive.

Telomere maps: Telomeres are the tips of the chromosomes. They are crucial in maintaining the chromosomes' stability and are important in the cell cycle and ageing. Because of the way the physical maps are constructed, many telomeres of chromosomes are left out.

Teratogens: Any agent that raises the incidence of congenital malformations.

Trait: Any detectable phenotypic property of an organism.

Transcript maps: In only a year or two, most human genes will be sequence-tagged and placed on various physical maps. Such a ï¿½transcript map' (or ï¿½expression map') of the genome will be an important part of the sequencing infrastructure, as well as a critical resource for the positional candidate approach to gene cloning. One of the specific goals of the US Human Genome Project is the construction of a high resolution STS map of the genome. .. One of the early problems with gene- based STSs was that there simply were not enough unique human gene sequences to bother with. But all of that changed with the advent of EST sequencing, at which time several groups began mapping ESTs albeit on a limited scale and only to the resolution of a chromosome assignment.

Transcription: The formation of an RNA molecule upon a DNA template by complementary base pairing.

Transcriptome maps: Consist of "expression clusters" of co-regulated genes. Challenges ahead for computational biology include the integration of clusters obtained for the transcriptome, the interactome, the phenome, and the localizome.

Transduction: The transfer of bacterial genetic material from one bacterium to another using a phage as a vector.

Transferase: Enzymes that catalyze the transfer of functional groups between donor and acceptor molecules.

Transgenic organism: One into which a cloned genetic material has been experimentally transferred, a subset of these foreign gene express themselves in their offspring.Turner syndrome a chromosomal condition in females (usually 45,XO) due to monosomy of the X- chromosome; characterized by short stature, failure to develop secondary sex characteristics, and infertility.

Translation: The formation of a polypeptide chain in the specific amino acid sequence directed by the genetic information carried by mRNA.

Translocation: A chromosome aberration which results in a change in position of a chromosomal segment within the genome, but does not change the total number of genes present.

Triplet code: A code in which a given amino acid is specified by a set of three nucleotides.

Tumor suppressor gene: genes that normally function to restrain the growth of tumors; the best understood case is for hereditary retinoblastoma.

UNESCO: United National Educational, Scientific, and Cultural Organization.

Vector: A self-replicating DNA molecule that transfers a DNA segment between host cells.

VNTR: Variable number tandem repeats; any gene whose alleles contain different numbers of tandemly repeated oligonucleotide sequences.

Von Hippel-Lindau syndrome: An autosomal dominant condition characterized by the anomalous growth and proliferation of blood vessels on the retina of the eye and the cerebellum of the brain; cysts and cancers in the kidneys, pancreas, and adrenal glands.

Western blotting analysis: A technique used to identify a specific protein; the probe is a radioactively labeled antibody raised against the protein in question.

Whole genome clone- based maps: In their paper, the International Human Genome Mapping Consortium describe how they constructed the first whole-genome physical map, how they created the templates from which the genome was sequenced and demonstrated how the map was essential for the accurate assembly of the human genome by the publicly funded effort. Four short reports accompanying the whole- genome mapping paper (Bruls; Bentley; Kucherlaparti; Page), describe alternative mapping strategies that were implemented for chromosomes 12, 14 and Y, as well as a host of other chromosomes. Information from all these papers were integrated into the whole- genome paper and demonstrate how a rich resource of mapping information can be generated by the cooperation of international independent efforts.

Whole proteome: Proteome analysis has become indispensable and complementary to genomic analysis. With access to whole genome sequences from various organisms and with the imminent completion of many more, the SWISS-PROT group at EBI has developed a research- oriented initiative that utilises many of the existing resources and provides comparative analysis of the predicted protein coding sequences of all complete genomes.

X-inactivation: The repression of one of the two X-chromosomes in the somatic cells of females as a method of dosage compensation; at an early embryonic stage in the normal female, one of the two X-chromosomes undergoes inactivation, apparently at random, from this point on all descendent cells will have the same X-chromosome inactivated as the cell from which they arose, thus a female is a mosaic composed of two types of cells, one which expresses only the paternal X-chromosome, and another which expresses only the maternal X-chromosome.

XYY syndrome: Genetic condition in males with extra Y chromosome (in 1 in 1000 male births). Symptoms: tall stature (over 6'), may including sterility, developmental delay, learning problems.

YAC maps: Yeast artificial chromosome maps, a type of physical map.

Yeast two hybrid: An approach to studying protein- protein interactions. The basic format involves the creation of two hybrid molecules, one in which a "bait" protein is fused with a transcription factor, and one in which a "prey" protein is fused with a related transcription factor. If the bait and prey proteins indeed interact, then the two factors fused to these two proteins are also brought into proximity with each other. As a result, a specific signal is produced, indicating an interaction has taken place.

Zoo blot: northern analysis of mRNA from different organisms.

Bibliography

A.J.F. Griffiths, J.H. Miller, D.T. Suzuki, R.C. Lewontin, and W.M. Gelbart (2000). An Introduction to Genetic Analysis. W.H. Freeman and Company. Chapter 2 (Patterns of Inheritance): Sex chromosomes and sex-linked inheritance.

Abreu, Eduardo L. et al. (2002). "Development of a program model to evaluate the potential for reuse of single-use medical devices: Results of a pilot test study". *Biomedical Instrumentation and Technology* 36.6, 89-404

Agaton C, Uhlen M, Hober S. (2004). Genome-based proteomics, *Electrophoresis*. 25(9):1280-8.

Agaton, C., J. Galli, I. Hoiden Guthenberg, L. Janzon, M. Hansson, A. Asplund, E. Brundell, S. Lindberg, I. Ruthberg, K. Wester, D. Wurtz, C. Hoog, J. Lundeberg, S. Stahl, F. Ponten, and M. Uhlen (2003). Affinity proteomics for systematic protein profiling of chromosome 21 gene products in human tissues. *Mol Cell Proteomics*, 2(6): p. 405-14.

Agaton, C., P. Unneberg, M. Sievertzon, A. Holmberg, M. Ehn, M. Larsson, J. Odeberg, M. Uhlen, and J. Lundeberg (2002)., Gene expression analysis by signature pyrosequencing. *Gene*, 289(1-2): p. 31-9.

Agris P (2004). "Decoding the genome: a modified view". Nucleic Acids Res 32 (1): 223–38.

Ahlstedt, S.A., and Bakaletz, S. (2005). *Assessment of freshwater mussel populations in Clear Creek, Tennessee (Emory-Obed River System) following an oil well fire and resulting spill* . Prepared for the National Park Service, Atlanta, Georgia.

Ahlstedt, S.A., Connell, J.F., Bakaletz, S., and Fagg, M.T. (2001). *Freshwater mussels of the National Park Service's Obed Wild and Scenic River, Tennessee* . Prepared for the National Park Service, Oneida, Tennessee.

Ahmadian, A., A. Russom, H. Andersson, M. Uhlen, G. Stemme, and P. Nilsson (2002). SNP analysis by allele-specific extension in a micromachined filter chamber. *Biotechniques*, 32(4): p. 748, 750, 752, 754.

Ahmadian, A., B. Gharizadeh, A.C. Gustafsson, F. Sterky, P. Nyren, M. Uhlen, and

J. Lundeberg (2000). Single-nucleotide polymorphism analysis by pyrosequencing. *Anal Biochem*, 280(1): p. 103-10.

Ahmadian, A., J. Lundeberg, P. Nyren, M. Uhlen, and M. Ronaghi (2000). Analysis of the p53 tumor suppressor gene by pyrosequencing. *Biotechniques*, 28(1): p. 140-4, 146-7.

Alberts B, Johnson A, Lewis J, Raff M, Roberts K, and Walter P (2002). Molecular Biology of the Cell, 4th ion. DNA and chromosomes: Chromosomal DNA and Its Packaging in the Chromatin Fiber.

Alberts B, Johnson A, Lewis J, Raff M, Roberts K, and Walter P (2002). Molecular Biology of the Cell, 4th ion. Proteins: The Shape and Structure of Proteins.

Alberts B, Johnson A, Lewis J, Raff M, Roberts K, and Walter P (2002). Molecular Biology of the Cell, 4th ion. Proteins: Protein Function.

Alberts B, Johnson A, Lewis J, Raff M, Roberts K, and Walter P (2002). Molecular Biology of the Cell, 4th ion. Control of Gene Expression - The Tryptophan Repressor Is a Simple Switch That Turns Genes On and Off in Bacteria.

Alberts, Bruce; Alexander Johnson, Julian Lewis, Martin Raff, Keith Roberts, and Peter Walters (2002). Molecular Biology of the Cell; Fourth Edition. New York and London: Garland Science.

Alberts, Bruce; Johnson, A, Lewis, J, Raff, M, Roberts, K & Walter, P (2002). *Molecular Biology of the Cell*, 4th edition, Garland.

Allemand, et al (1998). "Stretched and overwound DNA forms a Pauling-like structure with exposed bases". PNAS 24: 14152-14157.

Andersson A, Keskitalo J, Sjodin A, Bhalerao R, Sterky F, Wissel K, Tandre K, Aspeborg H, Moyle R, Ohmiya Y, Bhalerao R, Brunner A, Gustafsson P, Karlsson J, Lundeberg J, Nilsson O, Sandberg G, Strauss S, Sundberg B, Uhlen M, Jansson S, Nilsson P. (2004).A transcriptional timetable of autumn senescence. *Genome Biol*. 5(4):R24.

Andersson AC, Stromberg S, Backvall H, Kampf C, Uhlen M, Wester K, Ponten F. (2006). Analysis of protein expression in cell microarrays: a tool for antibody-based proteomics. *J Histochem Cytochem*. 54(12):1413-23.

Andrade, Berglund, Uhlen and Odeberg (2006). Using Grid Technology for Computationally Intensive Applied Bioinformatics Analyses, *In Silico Biology*, 495 - 504.

Astbury W, (1947). "Nucleic acid". Symp. SOC. Exp. Bbl 1 (66).

Ausubel, F.M. et al. (1989). *Current Protocols in Molecular Biology*. New York: John Wiley and Sons.

Ausubel, F.M. et al. (1989). *Short Protocols in Molecular Biology.* New York: John Wiley and Sons.

Avery OT, MacLeod CM, and McCarty M (1944). "Studies on the Chemical Nature of the Substance Inducing Transformation of Pneumococcal Types: Induction of Transformation by a Desoxyribonucleic Acid Fraction Isolated from Pneumococcus Type III". Journal of Experimental Medicine 79 (1): 137-58. 35th anniversary reprint available.

Bandmann, N., E. Collet, J. Leijen, M. Uhlen, A. Veide, and P.A. Nygren (2000). Genetic engineering of the Fusarium solani pisi lipase cutinase for enhanced partitioning in PEG-phosphate aqueous two-phase systems. *J Biotechnol*, 79(2): p. 161-72.

Barnhart, Benjamin J. (1989). "DOE Human Genome Program". Human Genome Quarterly 1: 1. Retrieved 2005-02-03.

Basu H, Feuerstein B, Zarling D, Shafer R, Marton L (1988). "Recognition of Z-RNA and Z-DNA determinants by polyamines in solution: experimental and theoretical studies". J Biomol Struct Dyn 6 (2): 299–309.

Bateson, W (1907). "The Progress of Genetic Research". Wilks, W (or) Report of the Third 1906 International Conference on Genetics: Hybridization (the cross-breeding of genera or species), the cross-breeding of varieties, and general plant breeding, London: Royal Horticultural Society.

Begon, Michael; Townsend, CR & Harper, JL (2005). *Ecology: From Individuals to Ecosystems*, 4th edition, Blackwell Publishing Limited.

Bengtsson S, Krogh M, Szigyarto CA, Uhlen M, Schedvins K, Silfversward C, Linder S, Auer G, Alaiya A, James P. (2007). Large-Scale Proteomics Analysis of Human Ovarian Cancer for Biomarkers. *J Proteome Res.* 6(4):1440-1450.

Benham C, Mielke S (2005). "DNA mechanics". Annu Rev Biomed Eng 7: 21–53.

Berns K (1990). "Parvovirus replication". Microbiol Rev 54 (3): 316–29.

Bickle T, Krüger D (1993). "Biology of DNA restriction". Microbiol Rev 57 (2): 434–50.

Blatt, Robin J.R. (1993). "The Human Genome Project" in association with MCET The Mass LearnPike, Cambridge, Massachusetts.

Bledsoe, J. Gary (2000). "Ethical discussions in an undergraduate tissue engineering course".

Blomstergren, A., D. O'Meara, M. Lukacs, M. Uhlen, and J. Lundeberg (2000). Cooperative oligonucleotides in purification of cycle sequencing products. *Biotechniques*, 29(2): p. 352-4, 356, 358 passim.

Brennan, Mark G. and Mark A. Tooley (2000). "Ethics and the Biomedical Engineer". *Engineering Science and Education Journal* 9.1, 5-7.

Brock, T. and Madigan, M (1992). "Gene manipulation and genetic engineering," In: Biology of Microorganisms 6th Edition. Englewood Cliffs, NJ: Prentice Hall, pp. 280-305.

Brock, T. and Madigan, M (1992). "Gene manipulation and genetic engineering," In: Biology of Microorganisms 6th Edition. Englewood Cliffs, NJ: Prentice Hall, pp. 280-305.

Brody, Eugene (1993). *Biomedical Technology and Human Rights*. Aldershot, England: UNESCO.

Bronzino, Joseph D., Ellen J. Flannery, and Maurice Wade (1990). "Legal and Ethical Issues in the Regulation and Development of Engineering Achievements in Medical Technology - Part I". *IEEE Engineering in Medicine and Biology* 9.1, 79-81.

Bud, R. (1989). "Janus-faced biotechnology: An historical perspective." Trends in Biotechnology. 7: 230-233.

Cadet J, Delatour T, Douki T, Gasparutto D, Pouget J, Ravanat J, Sauvaigo S (1999). "Hydroxyl radicals and DNA base damage". Mutat Res 424 (1–2): 9–21.

Calfee, John E. and Elizabeth DuPre (2006). *"The Emerging Market Dynamic of Targeted Therapeutics," Health Affairs*. 25(5): 1302-1308.

Campbell, Neil (2004). *Biology*, 7th edition, Benjamin-Cummings Publishing Company.

Casada, Mark E. and James A. DeShazer (1995). "Teaching Professionalism, Design, & Communications to Engineering Freshmen". *ASEE Annual Conference Proceedings*, v 1, "Investing in the Future",1381-1386.

Cathcart R, Schwiers E, Saul R, Ames B (1984). "Thymine glycol and thymidine glycol in human and rat urine: a possible assay for oxidative DNA damage". Proc Natl Acad Sci U S A 81 (18): 5633–7.

Chalikian T, Völker J, Plum G, Breslauer K (1999). "A more unified picture for the thermodynamics of nucleic acid duplex melting: a characterization by calorimetric and volumetric techniques". Proc Natl Acad Sci U S A 96 (14): 7853–8.

Champoux J (2001). "DNA topoisomerases: structure, function, and mechanism". Annu Rev Biochem 70: 369–413.

Choudhuri S (2004). "Gene Regulation and Molecular Toxicology". Toxicology Mechanisms and Methods 15 (1): 1-23.

Clark, J.M. and Switzer, R.L. (1977). Experimental Biochemistry. New York: W.H. Freeman and Company.

Clausen-Schaumann H, Rief M, Tolksdorf C, Gaub H (2000). "Mechanical stability of single DNA molecules". Biophys J 78 (4): 1997–2007.

Cook, R.B., Suter, II, G.W., and Sain, E.R. (1998). Ecological risk assessment in a large river-reservoir: 1. Introduction and background [Electronic version]. *Environmental Toxicology and Chemistry*, 18(4), 581-588.

Cook, S.B. (2006). *Quantification of injury to aquatic resources resulting from the Pryor oil spill and fire, Obed Wild and Scenic River, Task 1* (pdf file). Final Draft Report.

Cook, S.B., Russ, W.T., and Goodfred, D.W. (2005). *Emory River Watershed biological assessment* (pdf file). Quarterly Report submitted to the U.S. Fish and Wildlife Service and Tennessee Wildlife Resources Agency.

Cook, S.B., Russ, W.T., and Goodfred, D.W. (2005). Emory River Watershed biological assessment (pdf file). Quarterly Report submitted to the U.S. Fish and Wildlife Service and Tennessee Wildlife Resources Agency.

Cook, S.B., Russ, W.T., and Goodfred, D.W. (2006). *Biological survey of the Emory River Watershed, Tennessee* (pdf file). Final Report submitted to the U.S. Fish and Wildlife Service and Tennessee Wildlife Resources Agency.

Cram, Nicholas, John Wheeler, and Charles S. Lessard (1995). "Ethical Issues of Life-sustaining Technology". *IEEE Technology and Society Magazine* 14.1, 21-28.

Cristianini, N. and Hahn, M. (2006). *Introduction to Computational Genomics*, Cambridge University Press.

Curtis, H. and Barnes, N. (1989). Biology. New York: Worth Publishers.

Dahm R (2005). "Friedrich Miescher and the discovery of DNA". Dev Biol 278 (2): 274–88.

Dame RT (2005). "The role of nucleoid-associated proteins in the organization and compaction of bacterial chromatin". Mol. Microbiol. 56 (4): 858-70.

Daniels, A.U. (1992). "Ethics Education at the Engineering/Medicine Interface". *Journal of Investigative Surgery* 5.3, 209-18.

David Mount (2000), Bioinformatics, Sequence and Genome Analysis, pp. 2-3, Cold Spring Harbor Laboratory Press,

Davies J, Stanley J (1989). "Geminivirus genes and vectors". Trends Genet 5 (3): 77–81.

Davis, Michael (1992). "Codes of Ethics, Professions, and Conflict of Interest: A Case Study of an Emerging Profession, Clinical Engineering". *Professional Ethics* 1.1-2, 179-95.

DeLisi, Charles (2001). "Genomes: 15 Years Later A Perspective by Charles DeLisi, HGP Pioneer". Human Genome News 11: 3–4. Retrieved 2005-02-03.

Department of the Interior, Fish and Wildlife Service. (2002). *Federal Register* (pdf file). 67(188). Retrieved from http://www.fws.gov/policy/library/02fr61016.html

Department of the Interior, Fish and Wildlife Service. (2004). *Federal Register* (pdf file). 69(168). Retrieved from http://www.fws.gov/policy/library/04-19340.html

Dickman M, Ingleston S, Sedelnikova S, Rafferty J, Lloyd R, Grasby J, Hornby D (2002). "The RuvABC resolvasome". Eur J Biochem 269 (22): 5492–501.

Dixon, Dougal (1981). *After Man - A Zoology of the Future.* New York: St. Martin's Press.

Dodds, John H. (1985). *Plant Genetic Engineering.* New York: Cambridge University Press.

Doherty A, Suh S (2000). "Structural and mechanistic conservation in DNA ligases.". Nucleic Acids Res 28 (21): 4051–8.

Douki T, Reynaud-Angelin A, Cadet J, Sage E (2003). "Bipyrimidine photoproducts rather than oxidative lesions are the main type of DNA damage involved in the genotoxic effect of solar UVA radiation". Biochemistry 42 (30): 9221–6.

Dubcovsky J., Luo MC, Zhong GY, et al (1996). "Genetic map of diploid wheat, Triticum monococcum L., and its comparison with maps of Hordeum vulgare L". Genetics 143 (2): 983–99.

Eaton, Margaret (2004). *Ethics and the Business of Bioscience*. Stanford, CA: Stanford Business Book.

Ehn, M., P. Nilsson, M. Uhlen, and S. Hober (2001)., Overexpression, rapid isolation, and biochemical characterization of Escherichia coli single-stranded DNA-binding protein. *Protein Expr Purif*, 22(1): p. 120-7.

Eklund, M., L. Axelsson, M. Uhlen, and P.A. Nygren (2002). Anti-idiotypic protein domains selected from protein A-based affibody libraries. *Proteins*, 48(3): p. 454-62.

Ellis NA, Ciocci S, German J (2001). "Back mutation can produce phenotype reversion in Bloom syndrome somatic cells". Hum Genet 108 (2): 167-73.

Fee and Alstine (2006). "PEG-proteins: Reaction engineering and separation issues." Chemical Engineering Science 61(3): 924-939.

Fielder, J.H. (1992). "The Bioengineer's Obligations to Patients". *Journal of Investigative Surgery* 5.3, 201-8.

Fleischmann, R. D. et al. (1995). "Whole-genome random sequencing and assembly of Haemophilus influenzae Rd.". Science 269: 496"512.

Fredrikson, G.N., B. Gullstrand, J. Westberg, A.G. Sjoholm, M. Uhlen, and L. Truedsson (1998). Expression of properdin in complete and incomplete deficiency: normal in vitro synthesis by monocytes in two cases with properdin deficiency type II due to distinct mutations. *J Clin Immunol*, 18(4): p. 272-82.

Freeland S.J.; Wu T. and Keulmann N. (2003) The Case for an Error Minimizing Genetic Code. Orig Life Evol Biosph. 33(4-5), 457-77.

Friedberg, E.C. (1985). *DNA Repair*. New York: WH Freeman and Company.

Fruton, J.S. (1999). *Proteins, Enzymes, Genes: The Interplay of Chemistry and Biology*. Yale University Press.

Gharizadeh, B., M. Kaller, P. Nyren, A. Andersson, M. Uhlen, J. Lundeberg, and A. Ahmadian (2003). Viral and microbial genotyping by a combination of multiplex competitive hybridization and specific extension followed by hybridization to generic tag arrays. *Nucleic Acids Res*, 31(22): p. e146.

Ghosh A, Bansal M (2003). "A glossary of DNA structures from A to Z". Acta Crystallogr D Biol Crystallogr 59 (Pt 4): 620–6.

Goodfred, D.W. (2006). *Characterization of benthic macroinvertebrate communities within the Emory River Watershed, Tennessee* (pdf file). Master's Thesis. Tennessee Technological University: Cookeville.

Goodman, D.C. (1987). *From Farming to Biotechnology: A Theory of Agro-industrial Development*. Oxford: Blackwell.

Goss, S. (2006). The Emory River Watershed (including the Obed). *Tennessee Citizens for Wilderness Planning*, 268. Retrieved September 11, 2006, from http://www.tcwp.org/nl268.shtml

Graslund, S., M. Eklund, R. Falk, M. Uhlen, P.A. Nygren, and S. Stahl (2002). A novel affinity gene fusion system allowing protein A-based recovery of non-immunoglobulin gene products. *J Biotechnol*, 99(1): p. 41-50.

Graslund, S., M. Larsson, F. Sterky, M. Uhlen, J. Lundeberg, C. Hoog, and S. Stahl (1999). Recovery of upstream cDNA sequences by a PCR-based biotin-capture method. *Biotechniques*, 27(3): p. 488-91, 493-5, 497-8.

Graslund, S., M. Larsson, R. Falk, M. Uhlen, C. Hoog, and S. Stahl (2002). Single-vector three-frame expression systems for affinity-tagged proteins. *FEMS Microbiol Lett*, 215(1): p. 139-47.

Graslund, T., G. Lundin, M. Uhlen, P.A. Nygren, and S. Hobėr (2000). Charge engineering of a protein domain to allow efficient ion-exchange recovery. *Protein Eng*, 13(10): p. 703-9.

Graslund, T., M. Ehn, G. Lundin, M. Hedhammar, M. Uhlen, P.A. Nygren, and S. Hober (2002). Strategy for highly selective ion-exchange capture using a charge-polarizing fusion partner. *J Chromatogr A*, 942(1-2): p. 157-66.

Graslund, T., M. Hedhammar, M. Uhlen, P.A. Nygren, and S. Hober (2002). Integrated strategy for selective expanded bed ion-exchange adsorption and site-specific protein processing using gene fusion technology. *J Biotechnol*, 96(1): p. 93-102.

Green, James, D; Tsang, Lincoln; Cavagnaro, Joy (2003). "'Generic' or 'Follow-on' Biologics: Scientific Considerations and Safety Issues," *Expert Opin. Biol. Ther.*; 3(7):1019-1022.

Gregory T (2005). "The C-value enigma in plants and animals: a review of parallels and an appeal for partnership". Ann Bot (Lond) 95 (1): 133–46.

Griffiths AJF, Miller JH, Suzuki DT, Lewontin RC, and Gelbart WM (2000). An Introduction to Genetic Analysis. W.H. Freeman and Company. Chapter 2 (Patterns of Inheritance): Mendel's experiments

Grundfest, Warren S. and Andrea K. Scott (1998). "Ethical Issues in the Development of Medical Devices: The Role of Education and Training in the Evaluation of Safety and Efficacy". *Critical Reviews in Biomedical Engineering* 26.5-6, 378.

Guilbeau, Eric J. and Vincent B. Pizziconi (1998). "Increasing Student Awareness of Ethical, Social, Legal, and Economic Implications of Technology". *Journal of Engineering Education* 87.1, 35-44.

Gulich, S., M. Linhult, P. Nygren, M. Uhlen, and S. Hober (2000). Stability towards alkaline conditions can be engineered into a protein ligand. *J Biotechnol*, 80(2): p. 169-78.

Gulich, S., M. Uhlen, and S. Hober (2000).Protein engineering of an IgG-binding domain allows milder elution conditions during affinity chromatography. *J Biotechnol*, 76(2-3): p. 233-44.

Gustafsson, A.C., J.M. Kijas, A. Alderborn, M. Uhlen, L. Andersson, and J. Lundeberg (2001). Screening and scanning of single nucleotide polymorphisms in the pig melanocortin 1 receptor gene (MC1R) by pyrosequencing. *Anim Biotechnol*, 12(2): p. 145-53.

Haab BB, Paulovich AG, Anderson NL, Clark AM, Downing GJ, Hermjakob H, Labaer J, Uhlen M. (2006). A reagent resource to identify proteins and peptides of interest for the cancer community: a workshop report. *Mol Cell Proteomics*. 5(10):1996-2007.

Hedhammar M, Stenvall M, Lonneborg R, Nord O, Sjolin O, Brismar H, Uhlen M,Ottosson J, Hober S. (2005). "A novel flow cytometry-based method for analysis of expression levels in Escherichia coli, giving information about precipitated and soluble protein" *J Biotechnol*. 23;119(2):133-46

Heller, J.C. (2000). "Beyond a Code of Ethics for Bioengineers: The Role of Ethics in an Integrated Compliance Program". *Critical Reviews in Biomedical Engineering* 28.3-4, 507-11.

Hertzberg, M., H. Aspeborg, J. Schrader, A. Andersson, R. Erlandsson, K. Blomqvist, R. Bhalerao, M. Uhlen, T.T. Teeri, J. Lundeberg, B. Sundberg, P. Nilsson, and G. Sandberg (2001). A transcriptional roadmap to wood formation. *Proc Natl Acad Sci U S A*, 98(25): p. 14732-7.

Hjalmarsson S, Alderborn A, Fock C, Muldin I, Kling H, Uhlen M, Engstrand L. (2004). Rapid combined characterization of microorganism and host genotypes using a single technology. *Helicobacter*. 9(2):138-45.

Hoagland, Mahlon (2001). *The Way Life Works*, reprint edition, Jones and Bartlett Publishers inc.

Hober, S., J. Lundstrom Ljung, M. Uhlen, and B. Nilsson (1999). Insulin-like growth factors I and II are unable to form and maintain their native disulfides under in vivo redox conditions. *FEBS Lett*, 443(3): p. 271-6.

Holmberg A, Blomstergren A, Nord O, Lukacs M, Lundeberg J, Uhlen M. (2005). "The biotin-streptavidin interaction can be reversibly broken using water at elevated temperatures" *Electrophoresis*, 26(3):501-10

Holmberg K, Persson ML, Uhlen M, Odeberg J. (2005). "Pyrosequencing analysis of thrombosis-associated risk markers" *Clin Chem.*, 51(8):1549-52

Hubscher U, Maga G, Spadari S (2002). "Eukaryotic DNA polymerases". Annu Rev Biochem 71: 133–63.

Hüttenhofer A, Schattner P, Polacek N (2005). "Non-coding RNAs: hope or hype?". Trends Genet 21 (5): 289–97.

Isaksson J, Acharya S, Barman J, Cheruku P, Chattopadhyaya J (2004). "Single-stranded adenine-rich DNA and RNA retain structural characteristics of their respective double-stranded conformations and show directional differences in stacking pattern". Biochemistry 43 (51): 15996–6010.

Jansson, B., M. Uhlen, and P.A. Nygren (1998). All individual domains of staphylococcal protein A show Fab binding. *FEMS Immunol Med Microbiol*, 20(1): p. 69-78.

Jansson, M., G. Andersson, M. Uhlen, B. Nilsson, and J. Kordel (1998). The insulin-like growth factor (IGF)binding protein 1 binding epitope on IGF-I probed by heteronuclear NMR spectroscopy and mutational analysis. *J Biol Chem*, 273(38): p. 24701-7.

Jenuwein T, Allis C (2001). "Translating the histone code". Science 293 (5532): 1074–80.

Johansson T, Le Quere A, Ahren D, Soderstrom B, Erlandsson R, Lundeberg J, Uhlen M, Tunlid A.(2004). Transcriptional responses of Paxillus involutus and Betula pendula during formation of ectomycorrhizal root tissue. *Mol Plant Microbe Interact*. Feb;17(2):202-15.

Johansson, M.U., I.M. Frick, H. Nilsson, P.J. Kraulis, S. Hober, P. Jonasson, M. Linhult, P.A. Nygren, M. Uhlen, L. Bjorck, T. Drakenberg, S. Forsen, and M. Wikstrom (2002). Structure, specificity, and mode of interaction for bacterial albumin-binding modules. *J Biol Chem*, 277(10): p. 8114-20.

Johnson A, O'Donnell M (2005). "Cellular DNA replicases: components and dynamics at the replication fork". Annu Rev Biochem 74: 283–315.

Johnson, George B. (2005). *Biology, Visualizing Life*. Holt, Rinehart, and Winston.

Jonasson, P., P.A. Nygren, B.L. Johansson, J. Wahren, M. Uhlen, and S. Stahl (1998). Gene fragment polymerization gives increased yields of recombinant human proinsulin C-peptide. *Gene*, 210(2): p. 203-10.

Jonasson, P., P.A. Nygren, H. Jornvall, B.L. Johansson, J. Wahren, M. Uhlen, and S. Stahl (2000). Integrated bioprocess for production of human proinsulin C-peptide via heat release of an intracellular heptameric fusion protein. *J Biotechnol*, 76(2-3): p. 215-26.

Kwarteng, K.B. (2000). "Ethical Considerations for Biomedical Scientists and Engineers: Issues for the Rank and File". *Critical Reviews in Biomedical Engineering* 28. 3-4, 517-21.

Larsson K, Wester K, Nilsson P, Uhlen M, Hober S, Wernerus H. (2006). Multiplexed PrEST immunization for high-throughput affinity proteomics.*J Immunol Methods*. 315(1-2):110-20.

Larsson, M., S. Graslund, L. Yuan, E. Brundell, M. Uhlen, C. Hoog, and S. Stahl (2000). High-throughput protein expression of cDNA products as a tool in functional genomics. *J Biotechnol*, 80(2): p. 143-57.

Larsson, M., S. Stahl, M. Uhlen, and A. Wennborg (2000). Expression profile viewer (ExProView): a software tool for transcriptome analysis. *Genomics*, 63(3): p. 341-53.

Lerner M, Corcoran M, Cepeda D, Nielsen ML, Zubarev R, Ponten F, Uhlen M, Hober S, Grander D, Sangfelt O. (2007). The RBCC Gene RFP2 (Leu5) Encodes a Novel Transmembrane E3 Ubiquitin Ligase Involved in ERAD. *Mol Biol Cell*. 2007 Feb 21; [Epub ahead of print]

Levitan, M. (1988). Textbook of Human Genetics 3rd Edition. New York: Oxford Press. Luria, S.E. and Delbruck, M. (1943) "Mutations of Bacteria from Virus Sensitivity to Virus Resistance." Genetics. 28:491-511. Mangino, R.A. (1975) A Laboratory Manual for Microbiology. Portland, ME: J. Weston Walch.

Liljeqvist, S., F. Cano, T.N. Nguyen, M. Uhlen, A. Robert, and S. Stahl, Surface display of functional fibronectin-binding domains on Staphylococcus carnosus. *FEBS Lett*, 1999. 446(2-3): p. 299-304.

Linhult, M., H.K. Binz, M. Uhlen, and S. Hober (2002). Mutational analysis of the interaction between albumin-binding domain from streptococcal protein G and human serum albumin. *Protein Sci*, 11(2): p. 206-13.

Locatelli, Francesco; Aljama, Pedro; Barany, Peter; et al. (2004). "Erythropoiesis-Stimulating Agents and Antibody-Mediated Pure Red-Cell Aplasia: Where Are We Now and Where Do We Go From Here?," *Nephrol Dial Transplant*. 19: 288-293.

Lofblom J, Kronqvist N, Uhlen M, Stahl S, Wernerus H.(2007). Optimization of electroporation-mediated transformation: Staphylococcus carnosus as model organism. *J Appl Microbiol*. 2007, 102(3):736-47.

Lundberg E, Hoiden-Guthenberg I, Larsson B, Uhlen M, Graslund T. (2006). Site-specifically conjugated anti-HER2 Affibody molecules as one-step reagents for target expression analyses on cells and xenograft samples. *J Immunol Methods*. 2007 Jan 30;319(1-2):53-63.

Lundberg E, Sundberg M, Graslund T, Uhlen M, Andersson Svahn H. (2007). A novel method for reproducible fluorescent labeling of small amounts of antibodies on solid phase. *J Immunol Methods*. 2007 Feb 28; [Epub ahead of print]

Lynn, Richard; Van Court, Marilyn (2004). "New evidence of dysgenic fertility for intelligence in the United States". *Intelligence* 32 (2): p. 193. Ablex Pub.

Magner, LN (2002). *A History of the Life Sciences*. TF-CRC.

Micklos, D.A. and Freyer, G. A. (1990). DNA Science: A First Course in Recombinant DNA Technology. Cold Spring Harbor Laboratory Press and Carolina Biological Supply Company.

Micklos, D.A. and Freyer, G. A. (1990). DNA Science: A First Course in Recombinant DNA Technology. Cold Spring Harbor Laboratory Press and Carolina Biological Supply Company.

Morange, M & Cobb, M (2000). *A History of Molecular Biology*. Harvard University Press.

Morholt, E. and Brandwein, P.F. (1986) Sourcebook for the Biological Sciences 3rd Edition. San Diego: Harcourt Brace JOvanovich.

Mulcahy, Henry (1989). "A Simulated Laboratory Exercise in Genetic Engineering." Abstracts of the Annual Meeting of American Society for Microbiologists.

Mulcahy, Henry (1989). "A Simulated Laboratory Exercise in Genetic Engineering." Abstracts of the Annual Meeting of American Society for Microbiologists.

Mulder, Wernerus, Shi, Ponten, Hober, Uhlen and Hökfelt (2007). High-throughput screening of systematic generated antibodies against human gene products in the rat nervous system, *Neuroscience*, in press.

Mullis, Kary B. (1990). "The Unusual Origin of the Polymerase Chain Reaction." Scientific American. April: 56-61.

Naser, C.R. (2000). "What is Life, and What is a Machine? The Ontology of Bioengineering". *Critical Reviews in Biomedical Engineering* 28.3-4, 545-50.

National Park Service. (2005). *Obed National Wild and Scenic River* . Retrieved September 11, 2006, from http://www2.nature.nps.gov/geology/parks/obri/index.cfm

Naurato, N. and T.J. Smith (2003). "Ethical Considerations in Bioengineering Research". *Biomedical Sciences Instrumentation* 39: 573-8.

Nilsson, P., A. Larsson, J. Lundeberg, M. Uhlen, and P.A. Nygren (1999). Mutational scanning of PCR products by subtractive oligonucleotide hybridization analysis. *Biotechniques*, 26(2): p. 308-16.

Nilsson, P., D. O'Meara, F. Edebratt, B. Persson, M. Uhlen, J. Lundeberg, and P. Nygren (1999). Quantitative investigation of the modular primer effect for DNA and peptide nucleic acid hexamers. *Anal Biochem*, 269(1): p. 155-61.

Nord, K., O. Nord, M. Uhlen, B. Kelley, C. Ljungqvist, and P.A. Nygren (2001). Recombinant human factor VIII-specific affinity ligands selected from phage-displayed combinatorial libraries of protein A. *Eur J Biochem*, 268(15): p. 4269-77.

Nord, O., M. Uhlen, and P.A. Nygren (2003). Microbead display of proteins by cell-free expression of anchored DNA. *J Biotechnol*, 106(1): p. 1-13.

Nordstrom, T., K. Nourizad, M. Ronaghi, and P. Nyren (2000).Method enabling pyrosequencing on double-stranded DNA. *Anal Biochem*, 282(2): p. 186-93.

O'Meara, D., P. Nilsson, P.A. Nygren, M. Uhlen, and J. Lundeberg (1998). Capture of single-stranded DNA assisted by oligonucleotide modules. *Anal Biochem*, 255(2): p. 195-203.

Obed Wild and Scenic River, Tennessee (1978). *Development plan and stream classification* . Copy of report housed at the Fish and Wildlife Service office in Cookeville, Tennessee.

Odeberg, J., A. Ahmadian, C. Williams, M. Uhlen, F. Ponten, and J. Lundeberg (1999). Context-dependent Taq-polymerase-mediated nucleotide alterations, as revealed by direct sequencing of the ZNF189 gene: implications for mutation detection. *Gene*, 235(1-2): p. 103-9.

Odeberg, J., K. Holmberg, P. Eriksson, and M. Uhlen (2002). Molecular haplotyping by pyrosequencing. *Biotechniques*, 33(5): p. 1104, 1106, 1108.

Odeberg, J., O. Rosok, G.H. Gudmundsson, A. Ahmadian, L. Roshani, C. Williams, C. Larsson, F. Ponten, M. Uhlen, H.C. Asheim, and J. Lundeberg (1998). Cloning and characterization of ZNF189, a novel human Kruppel-like zinc finger gene localized to chromosome 9q22-q31. *Genomics*, 50(2): p. 213-21.

Old, R.W. and Primrose, S.B. (1989) *Principles of Gene Manipulation,* 4th Edition. London: Blackwell Scientific Publications.

Orchard S, Heck A, Uhlen M, Ping P. (2007). Publication Committee Meeting HUPO 5(th) Annual World Congress Long Beach, CA, USA 30 October 2006. *Proteomics*. 7(7):1009-11.

Parsons, J.W., and Benson, N.G. (1960). Fertilization of Obed River, Tennessee . *Journal of the Tennessee Academy of Science*, 3(1), 63-75.

Patrakka J, Xiao Z, Nukui M, Takemoto M, He L, Oddsson A, Perisic L, Kaukinen A, Szigyarto CA, Uhlen M, Jalanko H, Betsholtz C, Tryggvason K. (2007). Expression and subcellular distribution of novel glomerulus-associated proteins dendrin, ehd3, sh2d4a, plekhh2, and 2310066E14Rik. *J Am Soc Nephrol*. 18(3):689-97.

Pennington and Associates. (n.d.). *A survey of the Periphyton Community, Clear Creek, Tennessee* . For the National Park Service, Big South Fork National River and Recreation Area (BSFNRRA), Obed Wild and Scenic River.

Persson A, Hober S, Uhlen M. (2006). A human protein atlas based on antibody proteomics. *Curr Opin Mol Ther*. 8(3):185-90.

Persson, A., H. Backvall, F. Ponten, M. Uhlen, and J. Lundeberg (2002). Single cell gene mutation analysis using laser-assisted microdissection of tissue sections. *Methods Enzymol*, 356: p. 334-43.

Pettersson, B., A. Kodjo, M. Ronaghi, M. Uhlen, and T. Tonjum (1998). Phylogeny of the family Moraxellaceae by 16S rDNA sequence analysis, with special emphasis on differentiation of Moraxella species. *Int J Syst Bacteriol*, 48 Pt 1: p. 75-89.

Pettersson, B., G. Bolske, F. Thiaucourt, M. Uhlen, and K.E. Johansson (1998). Molecular evolution of Mycoplasma capricolum subsp. capripneumoniae strains, based on polymorphisms in the 16S rRNA genes. *J Bacteriol*, 180(9): p. 2350-8.

Pettersson, B., M. Wang, C. Fellstrom, M. Uhlen, G. Molin, B. Jeppsson, and S. Ahrne (2000). Phylogenetic evidence for novel and genetically different intestinal spirochetes resembling Brachyspira aalborgi in the mucosa of the human colon as revealed by 16S rDNA analysis. *Syst Appl Microbiol*, 23(3): p. 355-63.

Pettersson, B., S.K. de Silva, M. Uhlen, and F.G. Priest (2000)..Bacillus siralis sp. nov., a novel species from silage with a higher order structural attribute in the 16S rRNA genes. *Int J Syst Evol Microbiol*, 50 Pt 6: p. 2181-7.

Pienkowski, D. (2000). "The Need for a Professional Code of Ethics in Biomedical Engineering: A Lesson from History". *Critical Reviews in Biomedical Engineering* 28.3-4: 513-6.

Raleigh, R.F., and Nelson, P.C. (1985). *Habitat suitability, index models and instream flow suitability curves: pink salmon* (pdf file). Report no. 82(10.109). Fort Collins: Western Energy and Land Use Team, U.S. Fish and Wildlife Service.

Rasmussin and Matheson (eds) (1990). A Sourcebook of Biotechnology Activities. National Association of Biology Teachers, 11250 Roger Bacon Drive #19, Reston, VA 22090.

Rasmussin and Matheson (eds) (1990). A Sourcebook of Biotechnology Activities. National Association of Biology Teachers, 11250 Roger Bacon Drive #19, Reston, VA 22090.

Renberg B, Nordin J, Merca A, Uhlen M, Feldwisch J, Nygren PA, Karlstrom A.E. (2007). Affibody Molecules in Protein Capture Microarrays: Evaluation of Multidomain - Ligands and Different Detection Formats. *J Proteome Res.* 6(1):171-179.

Retzer, W.J. (1991). *Biotechnology Workbook.* Englewood Cliffs, NJ: Prentice Hall.

Riddle, J.W., Jr. (1975). *Status of the native muskellunge,* Esox masquinongy ohioensis, *of the Cumberland Plateau, Tennessee* . Master's Thesis. Tennessee Technological University: Cookeville.

Rigler, R., A. Pramanik, P. Jonasson, G. Kratz, O.T. Jansson, P. Nygren, S. Stahl, K. Ekberg, B. Johansson, S. Uhlen, M. Uhlen, H. Jornvall, and J. Wahren (1999).

Specific binding of proinsulin C-peptide to human cell membranes. *Proc Natl Acad Sci U S A.* 96(23): p. 13318-23.

Ronaghi, M., B. Pettersson, M. Uhlen, and P. Nyren (1998). PCR-introduced loop structure as primer in DNA sequencing. *Biotechniques*, 25(5): p. 876-8, 880-2, 884.

Ronaghi, M., M. Uhlen, and P. Nyren (1998).A sequencing method based on real-time pyrophosphate. *Science*, 281(5375): p. 363, 365.

Ronnmark, J., C. Kampf, A. Asplund, I. Hoiden-Guthenberg, K. Wester, F. Ponten, M. Uhlen, and P.A. Nygren (2003). Affibody-beta-galactosidase immunoconjugates produced as soluble fusion proteins in the Escherichia coli cytosol. *J Immunol Methods*, 281(1-2): p. 149-60.

Ronnmark, J., H. Gronlund, M. Uhlen, and P.A. Nygren (2002). Human immunoglobulin A (IgA)-specific ligands from combinatorial engineering of protein A. *Eur J Biochem*, 269(11): p. 2647-55.

Ronnmark, J., M. Hansson, T. Nguyen, M. Uhlen, A. Robert, S. Stahl, and P.A. Nygren (2002). Construction and characterization of affibody-Fc chimeras produced in Escherichia coli. *J Immunol Methods*, 261(1-2): p. 199-211.

Russ, W.T. (2006). *Current distribution and seasonal habitat use of the threatened spotfin chub in the Emory River Watershed* (pdf file). Master's Thesis. Tennessee Technological University: Cookeville.

Saha, Atanu; Grabowski, Henry; Birnbaum, Howard; Greenberg, Paul; Bizan, Oded (2006). "Generic Competition in the U.S. Pharmaceutical Industry," *International Journal of the Economics of Business*, 13(1): 15-38(24)

Sasser, M.A. (2003). *Indicator bacteria levels in the waters of two east Tennessee watersheds* . Master's Thesis. University of Tennessee: Knoxville.

Schellekens H. (2004). "Biosimilar Therapeutic Agents: Issues with Bioequivalence and Immunogenicity," *Eur J Clin Invest.* 34(12):797-799.

Schellekens, Huub (2002). "Bioequivalence and the Immunogenicity of Biopharmaceuticals," *Nature Reviews*; 1: 457-462

Schmalzer, P.A. (1982). *Vegetation of the Obed Wild and Scenic River, Tennessee, and a comparison of reciprocal averaging ordination and binary discriminant analysis* . Doctoral Dissertation. University of Tennessee: Knoxville.

Schwartz, Lewis B. (2002). "The Dilemma of Bioengineering Research on Human Subjects". Proceedings of the 2002 IEEE Engineering in Medicine and Biology 24th Annual Conference and the 2002 Fall Meeting of the Biomedical Engineering Society (BMES / EMBS), Oct 23-26 2002, Houston, TX. *Annual International*

Conference of the IEEE Engineering in Medicine and Biology - Proceedings v. 3: 2668-2669.

Schwenk JM, Lindberg J, Sundberg M, Uhlen M, Nilsson P. (2007). Determination of Binding Specificities in Highly Multiplexed Bead-based Assays for Antibody Proteomics. *Mol Cell Proteomics*. 6(1):125-132.

Slater, R.J. et al. (1986). *Experiments in Molecular Biology*. Clifton, NJ: Humana Press.

Smith, G.L., Mackett, M., and Ross, B. (1983). "Infectious vaccinia virus recombinants that express hepatitis B virus surface antigen." Nature . 302:490-495. Therman, E. and Susman, M. (1993) Human Chromosomes: Structures, Behavior and Effects 3rd Edition. New York: Springer-Verlag.

Smith, R.P. (1979). *Status report on the occurrence of the spotfin chub,* Hyboposis monacha (Cope)*, in the Emory River, Morgan County, Tennessee* (pdf file). Gatlinburg: Fishery Resources.

Spath, P.J., A.G. Sjoholm, G.N. Fredrikson, G. Misiano, R. Scherz, U.B. Schaad, B. Uhring-Lambert, G. Hauptmann, J. Westberg, M. Uhlen, C. Wadelius, and L. Truedsson (1999). Properdin deficiency in a large Swiss family: identification of a stop codon in the properdin gene, and association of meningococcal disease with lack of the IgG2 allotype marker G2m(n). *Clin Exp Immunol*, 118(2): p. 278-84.

Speir, H.J. (1969). *Comparison of rock bass,* Ambloplites rupestris (Rafinesque) *populations in Obed River and Spring Creek, Tennessee* . Master's Thesis. Tennessee Technological University: Cookeville.

Stedman, S., and Stedman, B. (2002). *Notes on the birds of the Big South Fork National River and Recreation Area and the Obed River National Wild and Scenic River* . Cookeville: Tennessee Technological University.

Stenvall M, Steen J, Uhlen M, Hober S, Ottosson J. (2005). "High-throughput solubility assay for purified recombinant protein immunogens" *Biochim Biophys Acta*., 31;1752(1):6-10

Sterky F, Bhalerao RR, Unneberg P, Segerman B, Nilsson P, Brunner AM, Charbonnel-Campaa L, Lindvall JJ, Tandre K, Strauss SH, Sundberg B, Gustafsson P, Uhlen M, Bhalerao RP, Nilsson O, Sandberg G, Karlsson J, Lundeberg J, Jansson S. (2004). A Populus EST resource for plant functional genomics. *Proc Natl Acad Sci U S A*. 101(38):13951-6.

Sterky, F., A. Holmberg, G. Alexandersson, J. Lundeberg, and M. Uhlen (1998). Direct sequencing of bacterial artificial chromosomes (BACs) and prokaryotic genomes by biotin-capture PCR. *J Biotechnol*, 60(1-2): p. 119-29.

Sterky, F., S. Regan, J. Karlsson, M. Hertzberg, A. Rohde, A. Holmberg, B. Amini, R. Bhalerao, M. Larsson, R. Villarroel, M. Van Montagu, G. Sandberg, O. Olsson, T.T. Teeri, W. Boerjan, P. Gustafsson, M. Uhlen, B. Sundberg, and J. Lundeberg (1998). Gene discovery in the wood-forming tissues of poplar: analysis of 5, 692 expressed sequence tags. *Proc Natl Acad Sci U S A*, 95(22): p. 13330-5.

Stirrat, G.M. (2003). "Education in Ethics". *Clinics in Perinatology* 30.1 :1-15.

Taussig MJ, Stoevesandt O, Borrebaeck CA, Bradbury AR, Cahill D, Cambillau C, de Daruvar A, Dubel S, Eichler J, Frank R, Gibson TJ, Gloriam D, Gold L, Herberg FW, Hermjakob H, Hoheisel JD, Joos TO, Kallioniemi O, Koegll M, Konthur Z, Korn B, Kremmer E, Krobitsch S, Landegren U, van der Maarel S, McCafferty J, Muyldermans S, Nygren PA, Palcy S, Pluckthun A, Polic B, Przybylski M, Saviranta P, Sawyer A, Sherman DJ, Skerra A, Templin M, Ueffing M, Uhlen M. (2007). ProteomeBinders: planning a European resource of affinity reagents for analysis of the human proteome.*Nature Methods* 4(1):13-7.

Tennessee Valley Authority (1970). *Tennessee valley streams: Their fish, bottom fauna, and aquatic habitat* (pdf file). *The Emory River, 1968*. Division of Forestry, Fisheries and Wildlife Development.

Torrey, J.G. (1987). "The Development of Plant Biotechnology." American Scientist. 73:354-363.

Towle, Albert (1988). *Modern Biology*. Austin, TX: Holt, Rinehart and Winston.

U.S. Department of the Interior. (1972). *Proposed Obed National Wild and Scenic River, Tennessee* . Draft Environmental Statement. Copy of report housed at the Fish and Wildlife Service office in Cookeville, Tennessee.

U.S. Department of the Interior. (1981). *Proposed Obed National Wild and Scenic River, Tennessee* . Draft Environmental Statement. Copy of report housed at the Fish and Wildlife Service office in Cookeville, Tennessee.

United States Department of Agriculture Natural Resources Conservation Service (2006). *WRP ranking appendix for 2006-2007* . Retrieved September 11, 2006, from http://www.tn.nrcs.usda.gov/programs/WRP/tande.html

Vener, T., A. Holmberg, J. Lundeberg, and M. Uhlen (1999). Selection analysis of single-stranded clones using solid-phase techniques. *Genet Anal,* 15(2): p. 47-50.

Vener, T., M. Nygren, A. Andersson, M. Uhlen, J. Albert, and J. Lundeberg (1998). Use of multiple competitors for quantification of human immunodeficiency virus type 1 RNA in plasma. *J Clin Microbiol*, 36(7): p. 1864-70.

Veronese and Pasut (2005). "PEGylation, successful approach to drug delivery." Drug Discovery Today 10(21): 1451-1458.

Vispé S, Defais M (1997). "Mammalian Rad51 protein: a RecA homologue with pleiotropic functions". Biochimie 79 (9-10): 587-92.

Waessmann, H., S. Zollner, A.C. Gustafsson, V. Wiebe, M. Laan, J. Lundeberg, M. Uhlen, and S. Paabo (2002). Extensive linkage disequilibrium in small human populations in Eurasia. *Am J Hum Genet*, 70(3): p. 673-85.

Wahl M, Sundaralingam M (1997). "Crystal structures of A-DNA duplexes". Biopolymers 44 (1): 45–63.

Walden, Richard (1988). *Genetic Transformation in Plants. Milton Keynes*, England: Open University Press.

Walker, G., Parisher, E., Smith, P., Whitlock, D., Kramer, D., Matthes, U., and Morefield, L. (n.d.). *Characterization of plant community structure and abiotic conditions on climbed and unclimbed cliff faces in the Obed River Gorge* . Copy of report housed at the Fish and Wildlife Service office in Cookeville, Tennessee.

Wang G, Vasquez KM (2006). "Non-B DNA structure-induced genetic instability". Mutat Res 598 (1–2): 103–119.

Wang J (2002). "Cellular roles of DNA topoisomerases: a molecular perspective". Nat Rev Mol Cell Biol 3 (6): 430–40.

Waterston RH, Lander ES, Sulston JE (2003). "More on the sequencing of the human genome". Proc Natl Acad Sci U S A. 100: 3022-4.

Wayne RK, Ostrander EA (1999). "Origin, genetic diversity, and genome structure of the domestic dog". Bioessays 21 (3): 247–57.

Westberg, J., A. Holmberg, M. Uhlen, and B. Pettersson (1999). Truncated fragments in polymerase chain reaction-based DNA sequencing. *Electrophoresis*, 20(3): p. 502-10.

Williams, C., F. Ponten, A. Ahmadian, Z.P. Ren, G. Ling, O. Rollman, A. Ljung, N.G. Jaspers, M. Uhlen, J. Lundeberg, and J. Ponten (1998). Clones of normal keratinocytes and a variety of simultaneously present epidermal neoplastic lesions contain a multitude of p53 gene mutations in a xeroderma pigmentosum patient. *Cancer Res*, 58(11): p. 2449-55.

Williams, C., F. Ponten, C. Moberg, P. Soderkvist, M. Uhlen, J. Ponten, G. Sitbon, and J. Lundeberg (1999). A high frequency of sequence alterations is due to formalin fixation of archival specimens. *Am J Pathol*, 155(5): p. 1467-71.

Williams, C., T. Norberg, A. Ahmadian, F. Ponten, J. Bergh, M. Inganas, J. Lundeberg, and M. Uhlen (1998). Assessment of sequence-based p53 gene analysis in human breast cancer: messenger RNA in comparison with genomic DNA targets. *Clin Chem*, 44(3): p. 455-62.

Williams, C.E., Bivens, R.D., and Carter, B.D. (2003). *A crayfish survey on Clear Creek, Morgan County, Tennessee* . Tennessee Wildlife Resources Agency.

Wirta V, Holmberg A, Lukacs M, Nilsson P, Hilson P, Uhlen M, Bhalerao RP, Lundeberg J. (2005). "Assembly of a gene sequence tag microarray by reversible biotin-streptavidin capture for transcript analysis of Arabidopsis thaliana" *BMC Biotechnol.*, 3;5(1):5

Wolfe, W.J. (2007). Geomorphology and vegetation of alluvial bars along the Obed Wild and Scenic River . *Proceedings of the Seventeenth Tennessee Water Resources Symposium*, Montgomery Bell State Park, Burns, Tennessee.

Wueste, Daniel E. (1997). "Professions, Professional Ethics, and Bioengineering". *Critical Reviews in Biomedical Engineering* 25.2: 127-49.

Watson J.D. and Crick F.H.C. "A Structure for Deoxyribose Nucleic Acid". (PDF) Nature 171, 737–738 (1953). Accessed 13 Feb 2007.

Wayne RK, Ostrander EA (1999). "Origin, genetic diversity, and genome structure of the domestic dog". Bioessays 21 (3): 247–57.

Wing R, Drew H, Takano T, Broka C, Tanaka S, Itakura K, Dickerson R (1980). "Crystal structure analysis of a complete turn of B-DNA". Nature 287 (5784): 755–8.

Wolfsberg T, McEntyre J, Schuler G (2001). "Guide to the draft human genome". Nature 409 (6822): 824–6.

Wray G (2002). "Dating branches on the tree of life using DNA". Genome Biol 3 (1): REVIEWS0001.

Wright W, Tesmer V, Huffman K, Levene S, Shay J (1997). "Normal human chromosomes have long G-rich telomeric overhangs at one end". Genes Dev 11 (21): 2801–9.

Yasukochi Y, Ashakumary LA, Baba K, Yoshido A, Sahara K (2006). "A second-generation integrated map of the silkworm reveals synteny and conserved gene order between lepidopteran insects". Genetics 173 (3): 1319–28.

Index